Ending Hepatitis C:
A Seven-step Plan for a Successful Eradication Program:
A Roadmap for Ending Endemic Disease Globally.

William A. Haseltine PhD
Kaelyn Varner, MPH

ACCESS Health Press

Recent Books by William A. Haseltine

Affordable Excellence: the Singapore Healthcare Story, William A Haseltine (2013)

Improving the Health of Mother and Child: Solutions from India; Priya Anant, Prabal Vikram Singh, Sofi Bergkvist, William A. Haseltine & Anita George (2014)

Modern Aging: A Practical Guide for Developers, Entrepreneurs, and Startups in the Silver Market, Edited by Sofia Widén, Stephanie Treschow, and William A. Haseltine (2015)

Aging with Dignity: Innovation and Challenge is Sweden-The Voice of Care Professionals; Sofia Widen and William A. Haseltine (2017)

Every Second Counts: Saving Two Million Lives. India's Emergency response System. The EMRI Story; William A Haseltine (2017)

Voices in Dementia Care; Anna Dirksen and William A Haseltine (2018)

Aging Well, Jean Galiana and William A. Haseltine (2019)

World Class. Adversity, Transformation and Success and NYU Langone Health, William A. Haseltine (2019)

Science as a Superpower: My Lifelong Fight Against Disease And The Heroes Who Made It Possible, William A. Haseltine (2021)

The Future of Medicine: Healing Yourself / Regenerative Medicine; William A Haseltine (2023)

Living ebooks

A Family Guide to Covid: Questions and Answers for Parents, Grandparents and Children, William A. Haseltine (2020)

A Covid Back To School Guide: Questions and Answers for Parents and Students, William A. Haseltine (2020)

Covid Commentaries: A Chronicle of a Plague, Volumes I, II, III, IV, V, and VI, William A. Haseltine (2020)

My Lifelong Fight Against Disease: From Polio and AIDS to Covid-19, William A. Haseltine (2020)

Variants!: The Shape-Shifting Challenge of Covid-19 Vaccine Evasion & Reinfection, William A. Haseltine (2021)

Covid Related Post-traumatic Stress Disorder (CV-PTSD): What It Is And What To Do About It, William A. Haseltine (2021)

Natural Immunity And Covid-19: What It Is And How It Can Save Your Life, William A. Haseltine (2022)

Omicron: From Pandemic to Endemic, William A. Haseltine (2022)

Monoclonal Antibodies: The Once and Future Cure for Covid-19, William A. Haseltine and Griffin McCombs (2023)

The Future of Medicine: Healing Yourself: Regenerative Medicine Part One, William A. Haseltine (2023)

Viroids and Virusoids: Nature's Own mRNAs, William A. Haseltine and Koloman Rath (2023)

Welcome to *Ending Hepatitis C: A Seven Step Plan for a successful eradication program: a roadmap for ending endemic disease globally.*

In 1980, the World Health Assembly declared smallpox the first infectious disease to be eradicated globally. As of 2022, polio is on track to be the second with poliovirus type two and type three already eradicated and poliovirus type one only present in Afghanistan and Pakistan. The elimination of these diseases has resulted from robust biomedical knowledge, strong global support, and heavily resourced and well-thought-out implementation strategies. With these same ideas, we now have an opportunity to add hepatitis C to that list. As one of the leading causes of liver disease in the world and a global burden of 58 million chronic infections, hepatitis C is a serious and persistent threat to public health. Even though we have resources to diagnose and cure the disease, it remains endemic in many countries and accounts for an estimated 290,000 deaths each year. So why is this disease still so prominent despite our ability to eradicate it?

As we delve into the pages of this book, we'll discuss many aspects of hepatitis C, including its history, transmission, impact on the body, available treatments, and the ongoing efforts to eliminate this disease. We will also discuss the successes of programs like Egypt's 100 Million Healthy Lives initiative and how those programs can be modified and scaled to be successful in countries of all economic levels.

Additionally, this book will serve as an educational tool and a source of inspiration and hope for eliminating other endemic diseases such as HIV, tuberculosis, and malaria. We have a tremendous opportunity to eliminate one of the biggest public health threats in the world and must only establish the will to do so.

The format of this book is something that I have dubbed a living ebook. I will continue to update *Ending Hepatitis C.* You may view these updates at no additional cost by visiting:

https://www.williamhaseltine.com/ending-hepatitis-c

Dedication

William Haseltine, PhD

To my wife, Maria Eugenia Maury, my children Mara and Alexander, my stepdaughters Karina, Manuela, and Camila, my three grandchildren Pedro, Enrique, and Carlos, and last but not least our three dogs, Sky, Luna, and Ginger.

Kaelyn Varner, MPH

To my parents, Ivra and LaMont, my siblings Karmyn and Carson, my confidants, Esther, Lex, Daly, and Theresa, and my dearest pug Barry.

Contents

Introduction

There are several infectious diseases including hepatitis B and C, HIV, malaria, and tuberculosis that are endemic in many places around the world. These diseases contribute to over 2.4 million global deaths annually but can be eliminated globally. We have already witnessed the successful eradication of diseases such as smallpox and polio through global screening and vaccination programs. Hepatitis C stands as another disease with the potential for worldwide elimination. There are an estimated 1.5 million new global hepatitis C cases each year, and the disease chronically infects over 58 million people globally. Given the substantial health consequences, economic impact, and potential for elimination, addressing hepatitis C is of paramount importance for public health organizations, policymakers, healthcare providers, and society. This book covers the seven essential steps necessary to accomplish that.

The Seven Steps to a Successful Disease Elimination Program

Step One: International Resolve

The first and most important step to eliminating these diseases is international resolve. The world must come together, recognize the burden of the disease, and commit to eliminating it. International resolve, however, must be backed by the global commitment of funds and resources. Elimination programs require research, drugs, and significant manpower. Several international institutions have committed themselves to providing those resources.

The World Bank is one of the top contributors to global health initiatives, and the funds they provide to borrowing countries support projects and programs that work to eliminate endemic diseases. As part of the Transforming Egypt's Healthcare System Project, the World Bank provided Egypt with a $530 million loan that was a key element of Egypt's hepatitis C elimination program. Like The World Bank, The Global Fund is another international organization focusing on funding programs addressing HIV/AIDS, tuberculosis, and malaria.

Effective and enduring disease elimination requires funding and long-term commitment by leadership and governments. Organizations like the Gates Foundation and World Health Organization are well-known for partnering with governments to provide low-cost treatment for population-wide initiatives. Recently, several organizations, including UNAIDS, PEPFAR, the Bill & Melinda Gates Foundation, and the Global Fund, formed a partnership with Indian pharmaceutical manufacturers that made TLD, a highly effective HIV treatment drug, affordable for low- and middle-income countries.

International resolve also involves a global commitment to funding research for drug and vaccine development. Gavi, the Vaccine Alliance, is a global organization that provides access to vaccines in low-income and middle-income countries. The Coalition for Epidemic Preparedness Innovations focuses on developing new and more effective vaccines. All these organizations play a crucial role in disease elimination.

International resolve, however, is just the first step. Following international resolve, the next critical steps involve increasing national awareness, expanding access to testing and treatment, and implementing preventive measures. From our analysis of successful national elimination programs, we have found the second most important requirement is the scientific means and biomedical tools necessary to detect and treat both those who are actively infected and those at risk. Next is the political will, popular support, and strong leadership needed for the meticulous implementation of those tools. This is critical for the following step which is the organization and the existence of a health system capable of diagnosing, treating, and following up with the entirety of a population. Finally, the cost of diagnostics and drugs must be affordable for both the government and the citizens. All these components are needed for a successful program and can be scaled and replicated to eliminate hepatitis C in any country.

Step Two: The Scientific Means and Biomedical Knowledge: The Hepatitis C Virus

Hepatitis C is caused by the hepatitis C virus: a single-stranded, positive sense, RNA virus in the Flaviviridae family. Although the virus is small, it can cause hefty consequences. Hepatitis C is the leading blood borne infection and the number one cause of liver disease in the United States. Acute infections with this virus are generally asymptomatic, but when left untreated, approximately up to 85% of those acute infections will develop into chronic infections. If allowed to progress, the blood borne viral infection can develop into liver disease which is marked by the inflammation and swelling of liver tissues. Once a case reaches chronic status, 20-30% will develop cirrhosis within 30 years of initial infection. For cases that progress to the cirrhosis stage: as much as 20% will develop cirrhosis-related liver failure and an estimated 1-4% of will progress to hepatocellular carcinoma.

Epidemiology

The incidence of new global hepatitis C infections has been steadily decreasing over the past few decades due to increased surveillance efforts and education campaigns risks and modes of transmission. Though these rates are decreasing, hepatitis C is still endemic in many regions of the world and thus remains a top global public health threat. In 2015, the World Health Organization found the global incidence of new hepatitis C infections to be around 1.75 million and has since set a target for an 80% decrease in new infections by 2030. New hepatitis infections and transmission also vary widely by world region. The high infection rate in the European (61.8 per 100,000) WHO region is most likely due to injection drug use. Likewise, the most common mode of transmission for new cases in the Eastern Mediterranean (62.5 per 100,000) WHO are due to unsafe needle practices in healthcare settings.

The prevalence of chronic hepatitis C presents a global burden of about 1% which translates to an estimated 71 million people and follows similar distribution patterns. The Eastern Mediterranean region has the greatest prevalence with 2.3% (15 million people). Followed by the European region, African region, and The Americas and Western Pacific. Within regions, China, Pakistan, India, Egypt, Russia, and the United States contribute to 50% of all global infections.

The Center for Disease Control estimates that 2.7 million people in the US are currently living with a chronic hepatitis C Virus infection with about 17,000 new chronic cases developing each year. The eastern and southeastern regions of the United States have the highest infection rates and incidence of both acute and chronic cases. Prevalence is also higher in rural regions, where public health resources are low, and treatment may be difficult to obtain. Baby boomers and millennials carry the most disease burden. Persons born between 1945 and 1965 have the highest prevalence and account for 75% of chronic cases. The high prevalence in this community can be attributed to high rates of intravenous drug use in the 1970s and 1980s and the absence of blood screening protocols. The rate of hepatitis C infections is highest among persons aged 20-29 and 30-39 years. The current spike in acute incidence rates among millennials aligns closely with the age cohort most affected by the opioid crisis. Prisons are notable hotspots for hepatitis C Virus transmission and mortality. Incarcerated individuals are at high-risk for hepatitis C, and an estimated 12-35% of people in prison have hepatitis C, a much larger percentage than the estimated 2% of the general US population.

Transmission and Risk Factors

Hepatitis C is transmitted through blood, and the virus can be present in even tiny amounts of blood and can survive outside the body for several weeks. Some groups are at a higher risk of contracting the disease than the general population. Injection drug users are among those at the most significant risk. Sharing needles or other equipment used for injecting drugs is one of the most common pathways for transmission in younger age cohorts. Additionally, sharing other contaminated equipment like tattoo needles, toothbrushes, and razors can also be pathways to infection. Before screening of the blood supply began in the United States in 1992, the hepatitis C virus could be transmitted through blood transfusions and organ transplants. Since the institution of blood supply screening in 1992, the current risk of transmission through this route is low but is a common reason for the high prevalence of hepatitis C among older age cohorts. Other avenues for transmission include mother-to-child infection during birth and high-risk sexual behavior such as having multiple partners. It is also important to note that having HIV or another sexually transmitted infection can increase the likelihood of acquiring hepatitis C. An estimated 21% of people with HIV in the United States have a hepatitis C coinfection. It is important to promote testing and transmission education among high-risk groups to prevent the progression of the disease to chronic stages and reduce the risk of long-term complications.

Diagnosis

Currently, diagnosing hepatitis C is a two step process. The initial step screens the blood for exposure to the hepatitis C virus. Whenever a person is infected with the virus, their bodies produce hepatitis C specific antibodies and exposure is determined by measuring the presence of hepatitis C antibodies in the blood. After confirmation that hepatitis C antibodies are present, an RNA test is performed to determine if there is active virus replication. This test is performed using a polymerase chain reaction to measure the levels of hepatitis C virus RNA in the blood. An additional PCR test may also be performed to determine the genotype of the virus, which plays an important role in deciding what treatment drugs to use. There are currently six different hepatitis C genotypes present across the globe. In the United States, genotypes 1, 2, and 3 make up over 98% of infections.

Immune Response

When acute hepatitis C infections are detected in the body, the body's innate immune response triggers a swift response. In the innate response, the immune system releases natural killer cells (NKs), proinflammatory cytokines, and the interferon-stimulated gene response (IGE) which work to interfere with HEPATITIS C replication and kill off infected hepatocytes. The innate response is often not enough to curb the replication of the hepatitis C virus in liver tissue but it allows the body time to mount an adaptive immune response. In the adaptive immune response, the body generates B-cells to induce the production of broadly neutralizing antibodies. These specialized antibodies are useful because they can recognize and block viral entry across multiple strains of hepatitis C. The immune system also releases two types of T-cells: helper CD4+ cells, which prompt B-cells to make antibodies, and cytotoxic CD8+ cells which kill cells already infected with the hepatitis C virus. Interestingly, up to 30% of hepatitis C infections will clear on their own. Research about these spontaneous clearance cases reveals the importance of having robust and sustained B-cell and T-cell responses. This research is especially useful as scientists work to develop an effective vaccine.

Drug Development

In the absence of a vaccine, hepatitis C is still highly treatable. The first generation of hepatitis C drugs focused on boosting the body's immune system response using interferon injections. These injections, however, were susceptible to disease relapse and came with many side effects. The next iteration of drugs relied on the pegylated alpha interferons in conjunction with ribavirin. The process of pegylation increases the molecule size of interferons which slows down the rate in which the drug was absorbed, thus increasing the amount of time the drug remained in the body. While this combination of drugs was a step up from interferon monotherapy, side effects persisted, and the efficacy of treatment varied greatly by genotype. In the early 2010s a new class of drugs called direct-acting antivirals, became available. Instead of boosting immune responses, these drugs were designed to attack specific molecular targets known to play a role in the replication of the hepatitis C virus. There are currently three classes of direct-acting antivirals: NS3/4A protease inhibitors that target the NS3/4A protease enzyme which is needed for the virus to develop the proteins necessary to replicate itself, NS5A inhibitors that work by blocking the NS5A multifunctional protein and inhibiting viral ability to assemble new virions, and NS5B inhibitors that block function of the NS5B membrane enzyme that initiates the synthesis of the hepatitis C virus's RNA genomes. Most direct-acting antiviral regimens consist of taking a daily pill for an average of 3 months with interval RNA testing performed to assess how well the treatment is working. Undetectable levels of viral RNA after completing treatment is considered a clinical cure, and at least 95% of patients who finish a course of direct-acting antiviral therapy achieve this status.

Vaccine Development

Although it is possible to design and execute a national program for the elimination of hepatitis C, there are other reasons for optimism in eradicating it. There is currently no vaccine for hepatitis C, but the ability to protect populations from getting the disease is important for curbing the incidence of new infections. One of the roadblocks to developing an effective vaccine is the hepatitis C virus' mutation ability. The virus has roughly 9600 nucleotides with 10^{-4} substitutions per site and round of replication which has led to existence of seven different genotypes and more than 67 subtypes. Globally, 75% of cases are genotype 1, 15 to 20% are genotype 2 or 3, and less than 5% are genotypes 4, 5, or 6. Despite the complications that the hepatitis C virus presents, the insight from spontaneous clearance cases and several other vaccines, like the pan-genotypic COVID-19 vaccine, have helped scientists develop promising methods for preventing infection from multiple strains. Research has shown that broadly neutralizing antibodies along with robust and sustained T-cell responses are required for clearing the disease. One promising method that scientists are using to induce production of T-cells and broadly neutralizing antibodies is by targeting conserved epitopes. An epitope is a specific part on the surface of a virus that T-cells can bind to and signal B-cells to produce hepatitis C antigen specific antibodies. Conserved epitopes are T-cell targets that appear across multiple strains of hepatitis C which would allow the vaccine to work against a breadth of antigen variants.

Step three: Political Will

A few countries, including Egypt and Georgia, have successfully eliminated hepatitis. Their programs offer valuable insight in how the elimination of this disease can be successful in any country, regardless of economic status. This can also be accomplished relatively inexpensively compared to other medical issues. If we look at the success of Egypt in particular, we can determine that the most important part of eliminating a disease within a country is the political will to do so. Up until the late 2010s, Egypt had the highest prevalence of hepatitis C in the world at 7% of the adult population.

Today, the country has been certified by the World Health Organization as virtually free of the disease. This is largely due to the action taken by President Abdel Fattah el-Sisi and the Egyptian government who decided to take swift and decisive action in response to the high prevalence. In September 2019, the Egyptian government implemented its 100 Million Healthy Lives program. It was funded by a $250 million loan from the World Bank, supported by President el-Sisi, and managed by the Ministry of Health and Dr. Hala Zaid. In less than a year, this program allowed Egypt to efficiently screen and treat its entire population over the age of 12 for hepatitis C.

In the past, the United States has lacked the political will to address the prevalence of hepatitis C but has recently taken first steps to change that through its 2023 National Hepatitis C Elimination Program. Over the course of 5 years and $11 billion the United States has developed a nationwide plan for promoting hepatitis C awareness and expanding access to testing and treatment, especially among vulnerable populations including those on Medicare, Medicaid, Indian Health Service, and incarcerated populations. Estimates put the 10 years savings at $18.1 billion which would continue to pay off as time goes by as efforts prevent downstream costs for liver transplants and chronic kidney disease.

Step Four: Popular Consent

Explicit government commitment is also a necessary part of a successful program. The popular consent of the population can generate both buy-in for scientists and encourage public health stakeholders to create new health policy and infrastructure. It also addresses social stigma and public health misconceptions. As seen with COVID-19 epidemic, public support can raise awareness about the disease, promoting prevention measures, and encouraging individuals to seek testing and treatment. There are many ways to generate popular consent. Egypt, for example, generated popular consent for the 100 Million Healthy Lives Program through a mass promotion campaign. Egypt's Ministry of Health flooded the country with advertisements for the campaign. Newspapers, TV and radio stations, social media platforms, and billboards and posters featuring President El-Sisi.

Step Five: An Effective Healthcare System

Once the political will and popular consent have been established, there must be a health system capable of reaching everybody in the country. Ideally, there would already be a central health system in place, but in some countries including the United States, the healthcare system is fragmented with many states or regions having their own infrastructure and care networks. These countries will have to establish a national system or create a standardized protocol to allow for an effective clinical cascade. In countries that do possess a central health system, it can enable a campaign to reach its entire population. Egypt's 100 Million Healthy Lives Program, for example, followed a simple cascade of following individuals from screening to post-treatment follow-up. The government started by launching an epidemiologic study to understand the scope of the issue in its target population of citizens 18 years and older. The next phase of the program was to educate the population about the goals and benefits of the program and encourage them to participate. In addition to a widespread promotional media campaign, citizens were encouraged to register for appointments on the centralized campaign website, www.stophcv.eg. This stimulated widespread interest to get tested at one of thousands of testing centers and mobile units stationed around the country. During their principal appointment, citizens were given an antibody test to test for viral exposure. Those with positive antibody results were then referred to specialized testing centers for PCR testing to confirm an active infection. Within 30 days of confirmed infection, the government was able to provide those individuals with three months of free treatment. Finally, citizens were retested at the end of their treatment regimens and, if found to be negative for infection, were provided a certificate of cure. Over the course of the program, 1.6 million adults tested positive for hepatitis C infections, 1.47 million accepted government provided treatment for no cost, and an estimated 1.23 million Egyptian citizens with active hepatitis C infections were cured.

An effective healthcare system is also crucial to streamlining the diagnostic process. As we know, current hepatitis C diagnosis requires two testing steps, which also requires two separate healthcare visits and leads to a lot of people being lost to follow-up. About a third of those tested for hepatitis C in the United States have incomplete testing which means they've had a reactive antibody test but haven't received an RNA test to confirm active infection. To address this, public health officials have proposed new requirements for automatic testing. These methods propose collecting two samples simultaneously: one for detecting antibodies and one for measuring RNA. Other proposals include using the same sample for both tests. A study of Veterans Affairs health facilities found that the current two-visit method resulted in 64% of veterans having complete testing and this increased to 98% when single visit testing was implemented. This testing is also more cost-effective.

Step 6: Low Cost Diagnostics and Treatment

Hepatitis C imposes a considerable economic burden on many countries. It affects workforce productivity and places a strain on healthcare resources. As a result, countries must be strategic about how they spend their funding, especially when it comes to treatment. For a large scale screening and treatment program to work, the cost of diagnostics and drugs must be affordable to both the government and the people. Lowering these costs can be accomplished through a variety of avenues including negotiations or patent agreements. Collaboration with international organizations like the World Health Organization and World Bank can help establish relationships between governments and pharmaceutical companies. Egypt, for instance, was able to leverage its status as a hepatitis C endemic country with the World Trade Organization to claim exemption for direct acting antiviral patents. The country was then free to purchase active pharmaceutical ingredients for seven different direct acting antivirals from India and formulate them by Egyptian pharmaceutical companies. These same drugs can retail for as high as $84,000 in some countries but Egypt's reduction in costs allowed the program to provide free treatment for all its citizens at a cost of just $45 USD to the Egyptian government. These efforts also allowed those who opted for private treatment to pay just $70 USD. If Egypt can successfully lower prices, there's no reason the United States and other high income countries should be paying 10-100 times more for the same tests and drugs. As public health leaders push for a single visit hepatitis C diagnostic, another aspect of the United States' recently announced National Hepatitis C Elimination Program supports the development of a single diagnostic test which would result in quicker diagnosis and fewer diagnostic costs. Currently, the World

Health Organization is in the process of approving the Xpert HCV VL FS test for use in low resource widespread elimination campaigns. This novel single diagnostic test can accurately detect active infection from a finger-stick sample in 1 hour. In addition to lowering diagnostic costs, many countries are putting forth plans to increase access to direct acting antivirals by lowering the cost of drugs. Another key aspect of the United States' 2023 National Hepatitis C Elimination Program supports the pharmaceutical subscription model which would substantially streamline and lower the total cost of testing and treatment. This model has been approved by a National Academies of Sciences, Engineering, and Medicine Committee. In this model, pharmaceutical companies would bid for a contract with the federal government which would allow the government unlimited access to drugs for certain populations in exchange for an upfront lump sum to the pharmaceutical company. In this model, both parties benefit as the government can drive down costs, treat more people, and save money on downstream health costs. The pharmaceutical companies recoup the costs of the drugs that many couldn't afford to purchase and use. This model was successfully pioneered in Louisiana and provided access to treatment for 11,100 residents all of which were Medicaid recipients or incarcerated in prisons. The subscription model also offers opportunities for expansion of treatment access to other populations such as those with diabetes.

Step Seven: Meticulous Execution

Finally, a successful eradication program requires meticulous execution. The 100 Million Healthy Lives campaign was fueled by the dedication of political and healthcare stakeholders at all levels. It also drew heavily from collaboration with both the medical community and general population. The government employed over 60,000 physicians, nurses, data analysts, and trained hundreds of volunteers, many of which had no previous medical background. The government also developed a virtual private network. It allowed officials to track the amount and age distribution of those screened to report in real time. The efficiency of the central organization system also allowed for the screening of other public health diseases in Egypt including hypertension, diabetes, and obesity. Other countries can streamline their implementation processes by incorporating primary care providers and telehealth platforms. This offers a cost effective way to educate the national population about harm reduction methods and the importance of testing. Additionally, telehealth enables specialists and primary care doctors to reach more isolated populations like those who live in rural areas. Although successful implementation of hepatitis C telehealth requires flexibility to optimize available resources, the widespread application of this method during the COVID-19 pandemic has demonstrated its effectiveness and feasibility across multiple aspects of medicine and healthcare.

Applications for Other Diseases

Many of the pillars of success that we discuss can set the tone for other disease elimination programs including those for HIV. In Sydney, Australia, widespread elimination efforts have positioned the city to be the first place in the world to meet the WHO's target of reducing the incidence of new HIV cases by 90%. According to a new study of HIV epidemiological trends between 2010 and 2022, Sydney reduced new infections among gay men by 88%. This success is widely due to the city's commitment to address its status as the epicenter of HIV by promoting a much higher rate of testing and use of pre-exposure prophylaxis (PrEP). Additionally, the city has expanded access to antiretroviral treatment which is now used by 95% of HIV-positive residents. Allowing access to these drugs has made the risk of transmission low. Similar measures have been used to address high incidences of HIV incidence in Botswana and South Africa. Botswana was the first African country to establish a national HIV treatment program and in 2022, Botswana also became the third country to reach the UNAIDS target of having 95% of all people living with HIV to be aware of their status with 95% of those aware of their status to be treated with sustained antiretroviral treatment (ART), and for 95% of people receiving ART achieving viral suppression. This feat was the result of a strong political will, cost collaboration, and strategic distribution of resources. In conjunction with the Center for Disease Control, the Botswana Ministry of Health first launched its Masa Program in 2002. Over the following decade, the program expanded HIV programming and established a "treat all" strategy to provide ART to all citizens living with HIV. When the program ended in 2015, Botswana continued its partnership with the CDC to focus on maintaining its programs and scaling up testing to prevent mother-to-child transmission and distribute prevention resources like PrEP.

Sri Lanka is another example of how these methods can be applied to eliminate other infectious diseases. Although the country has struggled with malaria control and resurgences due to several decades of internal conflict, in 2016, it was certified by the WHO as malaria free. They were able to accomplish this through sustained government dedication to vector control, treatment distribution, and disease surveillance. Like Egypt's hepatitis C campaign, Sri Lanka's Anti-Malaria Campaign depended on sustained commitment and collaboration between multiple public health stakeholders. First, the government established a strong disease surveillance program to quickly identify cases and detect spikes in incidence. The government was then able to swiftly treat those cases with newly available artemisinin-based combination therapy. In addition to using mobile vans for diagnosing and treating both their urban and rural populations, the government also partnered with local organizations around the country to distribute long-lasting insecticide-treated nets. Again, both the will of the government and the meticulous execution of an intervention program are paramount to elimination success.

In the quest to globally eliminate hepatitis, we know there are critical success factors for achieving this ambitious goal. Given the disease's substantial health consequences, economic impact, and potential for elimination, it is a definitive public health priority. As countries design programs to tackle the burden of hepatitis C within their borders, it is important to note that if those programs are deficient in any one of these: the political will, popular consent, an effective healthcare system, low prices for drugs and diagnostics, and meticulous implementation, they are unlikely to succeed. The commitment of government leaders is essential in driving initiatives and implementing policies necessary for large-scale screening, treatment, and prevention campaigns. Alongside political will, popular consent plays a pivotal role in generating public support, raising awareness, and breaking down social stigmas associated with hepatitis C. Countries must also create effective healthcare systems for by using centralized epidemiologic tracking systems, telehealth platforms, and innovative testing strategies to streamline diagnosis and treatment and reach vulnerable populations. Lowering the cost of diagnostics and treatment is another crucial aspect. Countries can use novel approaches like subscription models to make testing and treatment affordable for both governments and citizens. As we move forward in the journey to eliminate hepatitis C, we must remain committed to collaboration, innovation, and action. The lessons learned from Egypt's 100 Million Healthy Lives program can be applied to countries of all resource levels and have the potential to inform other disease eradication efforts. By following the roadmap outlined in this book and harnessing the power of biomedical science, political will, and healthcare infrastructure, we can move towards a future where hepatitis C is no longer a public health threat.

Hepatitis C, The Disease, Epidemiology, Treatment, Eradication Part 1: The Disease

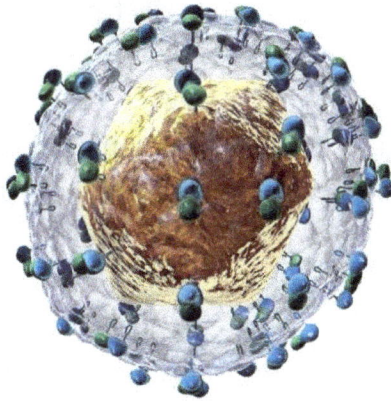

Hepatis C Virus (HCV)

Diagram of the structure of the hepatitis C virus particle.
Source: Blausen Medical

Hepatitis C is the leading cause of chronic liver disease and liver cancer in the world and affects an estimated 51 million people globally. Though the viral infection is easily diagnosable and treatable, a cure for hepatitis C remains elusive and one of the most prevalent public health issues in the United States and globally. In this article, we'll explain what hepatitis C is and how it affects the body, as well as explore the barriers to curing hepatitis and propose new strategies to eliminate the disease.

Discovery

The existence of what we now know to be hepatitis C was first discovered in 1975 by a team of NIH researchers led by Dr. Stephen Feinstone. While developing new serological testing to diagnose post-transfusion hepatitis, Feinstone's team found that many of their samples tested negative for both hepatitis A and hepatitis B, indicating that there was a new type of hepatitis caused by a novel infectious agent. This new, mysterious type of hepatitis was coined non-A, non-B hepatitis (NANBH) and was characterized by elevated liver transaminases closely associated with cirrhosis and liver disease. In 1978, Dr. Harvey J. Alter and his team at the NIH's Department of Transfusion Medicine were able to confirm the serial infectious ability of this new virus by injecting blood samples from NANBH positive humans into chimpanzees, which have a similar genetic makeup to humans. By studying the infected chimpanzees, Alter's team was able to understand the viral transmission of the new agent and learned that it was a small, enveloped virus. The identification of this new virus, however, proved difficult. The approaches used to identify the HAV and HBV viruses were unsuccessful in identifying HCV and necessitated the development of a new molecular cloning strategy. In 1985, George Kuo invented that cloning strategy by reverse transcribing mRNA from NANBH positive serum into cDNA. In 1989, a team at the Chiron Corporation led by Michael Houghton and Qui-Lim Choo were able to use that cDNA to successfully identify and replicate the virus. After publishing their findings, they named the new virus, the hepatitis C virus. HCV was then classified into the new genus of hepaciviruses in the family of Flaviviridae due to its similarities to the encephalitis flavivirus. In 2020 Dr. Alter and Haughton were awarded shares of the Nobel Prize in Physiology or Medicine for their part in discovering HCV.

Symptoms

Hepatitis C is a blood borne viral infection that affects the liver by causing inflammation and swelling of liver tissues. Acute infections are usually asymptomatic with only 20-30% of cases showing symptoms. When symptoms do occur, they present 2 weeks to 6 months after infection. Symptoms of an acute HCV infection can include jaundice (yellowing of the skin and eyes), dark-colored urine, light-colored stools, fatigue, abdominal pain, loss of appetite, nausea, diarrhea, joint pain, and fever.

Approximately 80% of those with acute hepatitis C will go on to develop a next stage, chronic infection. Chronic hepatitis is slow progressing and is often asymptomatic for decades after infection. The long interval between infection and presentation of symptoms often prevents chronic hepatitis C from being diagnosed until extensive liver damage has already occurred. As a result, chronic hepatitis C symptoms include cirrhosis symptoms like ascites (accumulation of fluid and swelling of the abdominal cavity), spider angioma on the stomach, jaundice, and easy bruising and bleeding in addition to symptoms also associated with an acute infection.

Extrahepatic symptoms can also manifest. Chronic hepatitis C has been associated with skin rashes including palpable purpura and digital ulcerations, as well as neuralgia and impaired cognitive function. Low blood pressure, congestive heart failure, and difficulty regulating blood sugar have also been linked to chronic hepatitis C.

Immune Reactions

Acute HCV infection is easily detected in the body and readily triggers the body's innate immune response. This first line defense against HCV employs the use of natural killer cells (NKs), proinflammatory cytokines, and the interferon-stimulated gene response (IGE) to fight the infection. The NKs and ISG responses work to interfere with HCV replication and kill off infected hepatocytes. While this swift, innate response is often not enough to curb the replication of HCV in liver tissue, it allows the body time to mount an adaptive immune response, a process that can take several weeks to months to develop. When the adaptive response starts taking effect, CD4+ and CD8+ T-cells are released which play an important role in coordinating a sustained immune response and killing off infected cells.

Spontaneous Clearance

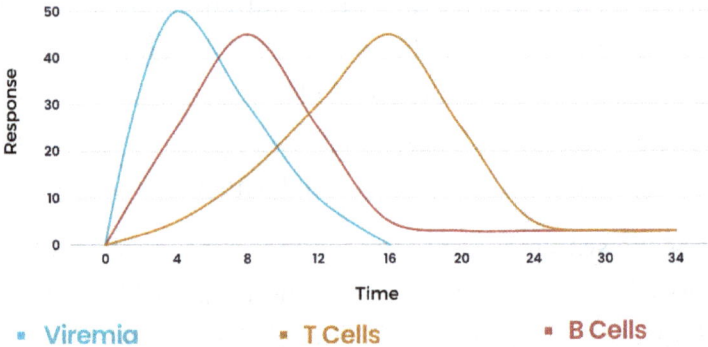

Chronic Infection

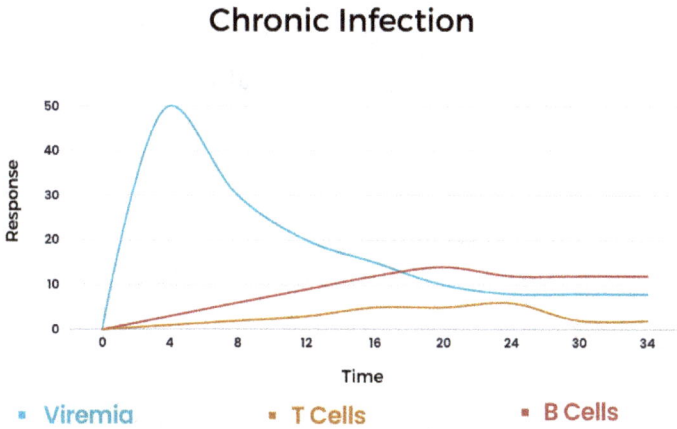

Figure 1. Immune reaction profile of (A) spontaneous clearance of HCV infection and (B) persistent HCV infection.

The profile and timing of the immune response to HCV can vary due to the capability of each body's immune system and genetic makeup. In spontaneous clearance of HCV infection, the innate immune response displays high ISG activity within the first few days of infection. As the ISG response successfully stops replication of HCV, there is a sharp decline in viremia around 8-12 weeks after infection (Figure 1). This sharp decline is also associated with the rise of CD4+ and CD8+ cells as well as the release of B-cell derived neutralizing antibodies. As the infection clears, the adaptive immune system response declines but can rapidly reemerge with reinfection. Research shows that broadly employed, rapid, and persistent CD8+ and CD4+ T-cell responses are critical in HCV clearance. In cases where infection persists, the ISG response remains high but is unable to adequately control HCV replication. A strong ISG response that coincides with high, but stable viremia is characteristic of a chronic infection. In these cases, the release of

CD4+ and CD+ T-cells is also delayed and less robust. These cells gradually exhaust and then disappear from blood circulation. The release of neutralizing antibodies is also delayed in chronic infections but they may persist longer than in spontaneous clearance cases.

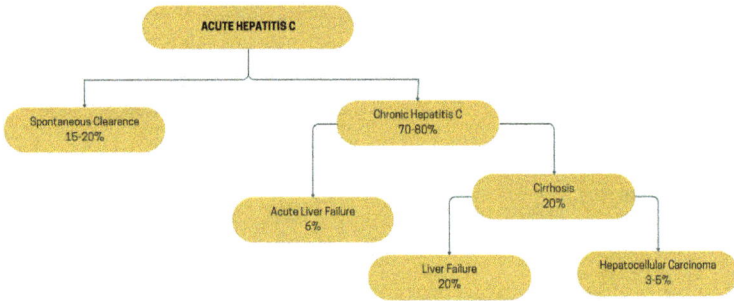

Figure 2. Natural history of a hepatitis C infection.

Long Term Consequences

There are many long-term consequences of a persistent hepatitis C infection. Most notably, chronic hepatitis C is associated with an increased risk of cirrhosis, liver disease, and liver cancer. After acquiring an acute hepatitis C infection, around 15 -20% of cases will undergo spontaneous clearance (Figure 2). The remaining 70-80% will go on to develop a persistent infection leading to chronic hepatitis C. Of those chronic cases: 6% will experience acute liver failure and up to 20% will develop cirrhosis within 10-15 years of initial infection. For cases that progress to the cirrhosis stage: 20% will develop cirrhosis-related liver failure within 20 years. Within a 30-year period, an estimated 3-5% of cirrhotic liver cases will progress to hepatocellular carcinoma (HCC), an aggressive cancer with an overall 5-year survival rate of 20%. In 2020, the CDC estimated that 1 in 5 people will die from chronic hepatitis c due to complications from cirrhosis and HCC.

Hepatitis C, The Disease, Epidemiology, Treatment, Eradication Part 2: Global Epidemiology

~~~~

The World Health Organization currently estimates the global burden of Hepatitis C to be between 130 and 170 million people. Ideally, this figure, as well as stratifications by location, age, and gender, would be more precisely determined based on data from community-based studies. In most countries, however, the utilization of such surveys is lacking and often only accounts for adults in high-risk groups such as blood donors and intravenous drug users instead of the general population. As global efforts to increase HCV surveillance become more prominent, predictive modeling offers valuable insight to both HCV incidence and prevalence around the world. In this chapter, we'll cover current estimates of HCV burden by geographic location, genotype, age, and gender as well as the global burden of HCV morbidity and mortality.

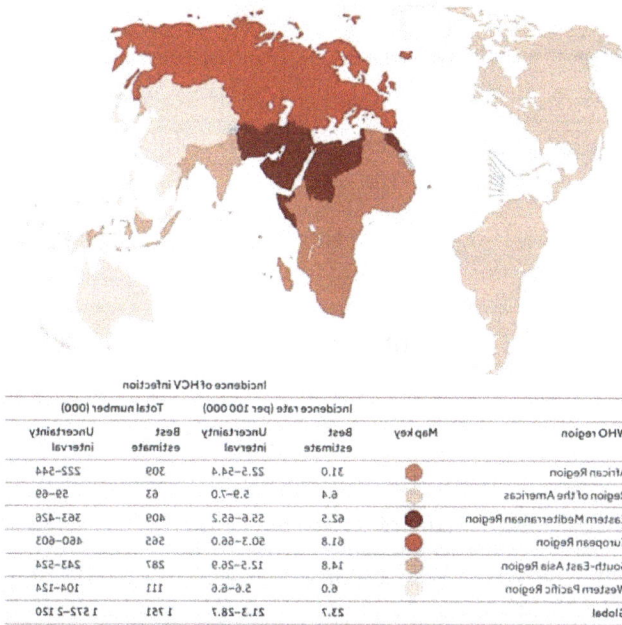

WHO region	Map key	Incidence rate (per 100 000)		Total number (000)	
		Best estimate	Uncertainty interval	Best estimate	Uncertainty interval
African Region		31.0	22.5–54.4	309	225–544
Region of the Americas		6.4	5.9–7.0	63	59–69
Eastern Mediterranean Region		62.5	55.6–69.2	409	365–426
European Region		61.8	50.4–66.0	565	460–603
South-East Asia Region		14.8	12.5–25.9	587	243–524
Western Pacific Region		6.0	3.6–6.6	111	104–124
Global		23.7	21.3–28.7	1 751	1 572–5 150

Figure 1. 2015 Incidence of hepatitis C in the general global population displayed by WHO region. Source: World Health Organization

Incidence and Prevalence by Geographic Region and Genotype

The incidence of new global HCV infections has been steadily decreasing in the past 50 years due to increased surveillance and education about HCV risks and transmission. Though these rates are decreasing, hepatitis C remains endemic in many countries. In 2015, the WHO found the global incidence of new HCV infections to be around 1.75 million and has since set a goal of reducing new HCV infections by 80% between 2015 and 2030.

HCV infections can be found in all regions of the world with the highest number of new infections occurring in the European (61.8 per 100,000) and Eastern Mediterranean (62.5 per 100,00) regions (Figure 1). Most new infections in the Eastern Mediterranean are due to unsafe needle practices in healthcare, while injection drug use is the most common mode of transmission in the European region. Following those regions, the African region (31 per 100,000) and South-East Asian region (14.8 per 100,000) have the next highest rates of infections. The lowest incidence rates are in the Americas (6.4 per 100,000) and the Western Pacific region (6.0 per 100,000).

Chronic hepatitis C presents a global burden of about 1% which translates to an estimated 71 million people. Similarly to global incidence, the prevalence of chronic HCV infections is also unevenly distributed around the globe with variations occurring both between and within regions (Figure 2). The Eastern Mediterranean region has the greatest prevalence with 2.3% (15 million people). Followed by the European region with 1.5% (14 million people) and African region with 1% (11 million people). The Americas and Western Pacific both have a 0.7% prevalence with 7 million and 14 million cases respectively. South-East Asian has the lowest prevalence at 0.5% (10 million people). Within regions, China, Pakistan, India, Egypt, Russia, and the United States contributed to 50% of all global infections.

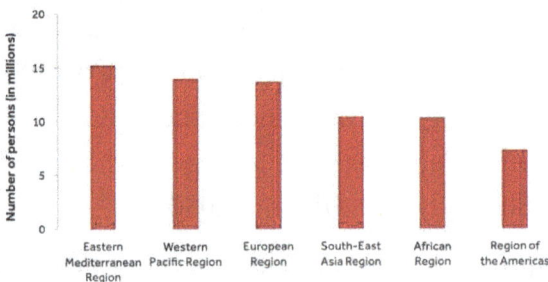

Figure 2. Number of persons with hepatitis C by WHO region in 2015. Source: World Health Organization

Prevalence by HCV genotype, which can affect the efficacy of hepatitis C treatment, also varies among different geographic regions. Of the 7 different genotypes, type 1 is the most common and accounts for 46% of global infections. It is the most common genotype found in Australasia, Europe, Latin America, and North America. HCV type 3 is the next most common and accounts for 22% of global cases. Genotypes 2 and 4 account for about 13% of global cases each with type 2 comprising an estimated 40% of cases in Asia and type 4 being especially prevalent in the North Africa and Middle East regions.

Incidence and Prevalence by Age and Sex

Though epidemiological data about hepatitis C by age and sex is sparse, studies have shown that incidence and prevalence vary by mode of transmission. Incidence of IV drug contracted HCV, the main mode of transmission in higher income countries, is highest among young men in these regions. Modeling also shows increasing incidence among women of childbearing age due to increased rates of IV drug use. Studies have also found the prevalence of medically contracted HCV infection to be higher among older populations, particularly those born between 1945 and 1965. This applies to most geographic regions and is possibly due to the cohort effect of being born before the implementation of blood screening.

Morbidity and Mortality

Although incidence rates are decreasing, mortality and morbidity continue to rise as older populations with chronic infections age and progress to more advanced stages of liver disease. In 2019, consequences of long-term HCV infection resulted in an estimated 290,000 deaths, most of which can be attributed to the development of liver cirrhosis and hepatocellular carcinoma. These complications also contribute to high rates of morbidity. In 2019 hepatitis C resulted in 15.3 million global Disability-adjusted life years (DALYs) and 0.6% of total global DALYs. Of those DALYs, 79.5% can be attributed to cirrhosis complications, and 18.9% can be attributed to liver cancer complications.

As global efforts to reduce the incidence, prevalence, and mortality of hepatitis C are becoming more widespread, better epidemiological surveillance is still needed to present a more accurate picture of hepatitis C burden and better inform strategies to eradicate the disease.

Hepatitis C, The Disease, Epidemiology, Treatment, Eradication Part 3: United States Epidemiology

※

Hepatitis C is the leading blood borne infection, and the number one cause of liver disease in the United States. The Center for Disease Control estimates that 2.7 million people in the US are currently living with a chronic Hepatitis C Virus infection with about 17,000 new chronic cases developing each year. The incidence of acute hepatitis C is equally concerning and has more than doubled since 2013. Though the national burden of hepatitis C presents an urgent public health emergency on its own, stratified data about its incidence, prevalence, and mortality show that the burden of hepatitis C varies by geographic and demographic groups. In this chapter, we'll illustrate the scope of hepatitis C infections in the United States and highlight the inequities in disease burden across various groups of people.

Surveillance of Hepatitis C Virus in the United States

Epidemiological data on hepatitis C is dependent on data reported by individual health departments to the CDC. Since different states and jurisdictions have variable access to surveillance resources, estimates of incidence and prevalence are often greater than reported cases. In 2020, a total of 4,798 acute infections were reported to the CDC, but extrapolated estimates put the total number of acute infections to be around 66,700 cases. The CDC has instituted several strategies to address underreporting and promote more accurate data collection. CDC criteria of acute hepatitis C cases has become increasingly sensitive in recent years, and the specificity has enabled health departments to better catch cases that would not have been classified as hepatitis C in the past. Increasing testing efforts have also enabled smaller health departments to catch acute cases which are most often asymptomatic and less likely to be diagnosed.

Disease Burden by Geographic Region

The incidence of both acute and chronic hepatitis C infection varies by region of the United States. In 2019, the national rate of acute infection was 1.3 cases per 100,000 people. When stratified by state, however, the eastern and southeastern regions had the highest infection rates. In 2020, Maine had the highest incidence (11.9 cases per 100,000), followed by Florida (6.1 cases per 100,000), Louisiana (6.0 cases per 100,000), and West Virginia (5.3 cases per 100,000) (Figure 1). In that same year, the states with the highest incidence of chronic hepatitis C were in West Virginia (122.1 cases per 100,000 population), followed by Alabama (115.8 cases per 100,000 population), Louisiana (90.6 cases per 100,000 population), and Mississippi (88.1 cases per 100,000 population). States that have been hit hardest by the opioid epidemic tend to have the highest rates of new infection, and as the opioid crisis continues to grow, those rates are rapidly increasing. From 2006 to 2012, acute hepatitis C infections increased by 364% in Kentucky, Tennessee, Virginia, and West Virginia. Incidence of both acute and chronic cases is also higher in rural regions, where public health resources are low, and treatment may be difficult to obtain.

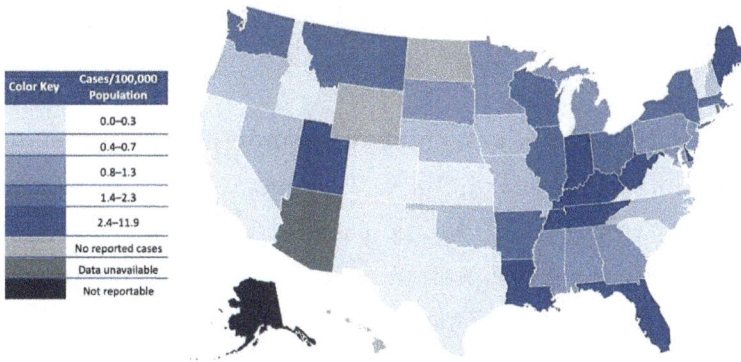

Figure 1. Map of acute hepatitis C virus infection incidences in 2020. Maine had the highest rate of 11.9 new cases per 100,000 people. Source: Center for Disease Control

Disease Burden by Demographic Characteristics

Increased surveillance in the United States has also revealed demographic differences in hepatitis C epidemiology. Historically, males have had higher rates for chronic infection and of the 107,300 newly chronic cases reported in 2020, 64% occurred in males. In 2019, the incidence of acute hepatitis C infections was also slightly higher for males (1.6 new infections per 100,000) than females (1.0 new infections per 100,000) (Figure 2).

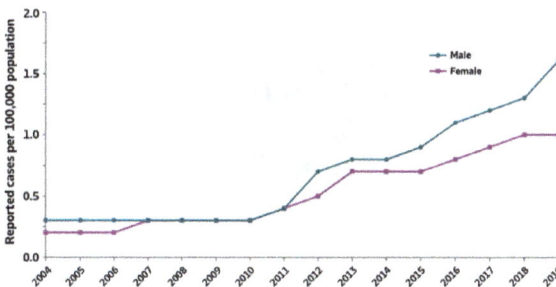

Figure 2. Incidence of acute hepatitis C virus infections by sex from 2005 to 2020. The 2020 incidence of acute hepatitis C infections was higher for males

(2.0 new infections per 100,000) than females (1.0 new infections per 100,000).
Source: Center for Disease Control

Baby boomers and millennials carry the most disease burden when compared to other age cohorts. Older generations have the highest prevalence with 75% of chronic cases appearing in persons born between 1945 and 1965, prompting many hospitals and healthcare institutions to implement mandatory Hepatitis C Virus testing for persons in that cohort. In 2019, both the reported number and rate of acute hepatitis C infections were highest among persons aged 20-29 and 30-39 years of age (Figure 3). It is important to note that the chronic infection prevalence among persons born between 1945 - 1965 corresponds with the high incidence rate that occurred among young adults using intravenous drugs in the 1970s and 1980s. Similarly, the current spike in acute incidence rates among millennials aligns closely with the age cohort most affected by the opioid crisis. A greater focus on limiting transmission through injection drug use is, therefore, instrumental in addressing the rising incidence of acute cases in the United States.

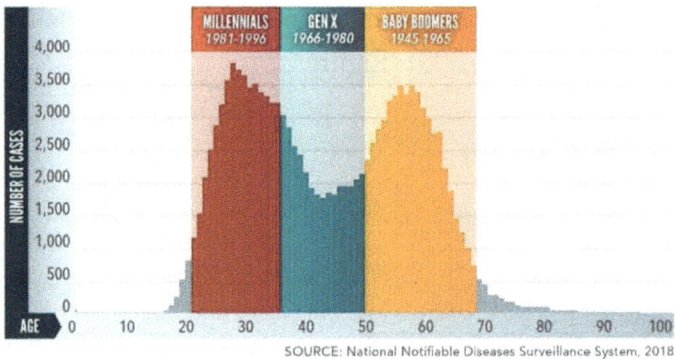

SOURCE: National Notifiable Diseases Surveillance System, 2018

Figure 3. 2018 hepatitis C prevalence by age cohort. Millennials and baby boomers have the highest numbers of infection. Source: National Notifiable Disease Surveillance System

American Indian/Alaskan Native and non-Hispanic black communities are disproportionately affected by the Hepatitis C Virus (Figure 4). The incidence of acute infections in 2020 was highest among those of American Indian/Alaskan Native ethnicity (1.2 per 100,000) followed by white non-Hispanic (1.6 per 100,000), black non-Hispanic (1.1 per 100,000), Hispanic ethnicity (0.7 per 100,000), and Asian/Pacific Islander (0.4 per 100,000). Trend data collected from 2003 to 2019, shows a dramatic increase in the rate of reported acute infections among American Indian/Alaska Native individuals while data from 1999-2016 shows the highest hepatitis C prevalence among Black persons.

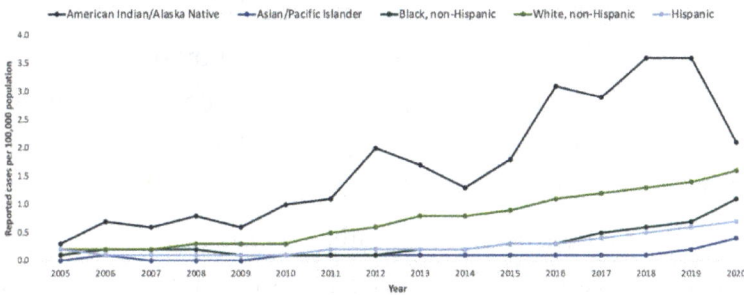

Figure 4. Incidence of acute hepatitis C virus infections by race/ethnicity from 2005 to 2020. The incidence of acute infections in 2020 was highest among those of American Indian/Alaskan Native ethnicity (1.2 per 100,000). Source: Center for Disease Control

Disease Burden Among the US Incarcerated Population

Although prisons and correctional facilities are required to treat and care for the medical needs of their populations, they are notable hotspots for Hepatitis C Virus transmission and mortality. Many incarcerated individuals are at high-risk for hepatitis C due to previous intravenous drug use and unsanitary tattoo needle use while incarcerated. It's estimated that between 12% and 35% of people in prison have hepatitis C, a much larger percentage than the estimated 2% of the general US population. Without widespread testing and access to curative treatments, incarcerated individuals also carry a disproportionate mortality burden. In 2019, the hepatitis C-related death rate for incarcerated individuals was more than double the rate in the general population. These inequities require immediate attention to address preventative mortality and morbidity.

Morbidity and Mortality

In 2019, there were 15,000 hepatitis C related deaths in the United States—a mortality rate of 3.33 deaths per 100,000 people. Though this rate was a 32% decrease from the 2015 mortality rate, not all geographic and demographic groups have seen such a decline. States with the highest mortality rates in 2019 were Oklahoma and the District of Columbia (10.75 and 10.08 deaths per 100,000 respectively). California (2,114 deaths), Texas (1,383 deaths), and Florida (1,025 deaths) had the highest number of reported hepatitis C-associated deaths and accounted for more than 30% of all the hepatitis C related deaths reported in 2019. In 2020, the total number of hepatitis c-related deaths was highest in White persons (9,397 people), but death rates were highest among American Indian/Alaskan Native and non-Hispanic Black persons (10.17 per 100,000 and 5.63 times per 100,000, respectively) (Figure 5). In fact, Hepatitis C Virus related chronic liver disease is currently the 5th leading cause of death among American Indian/Alaskan Native persons. Although men experience the highest burden of Hepatitis C related death, the rates are increasing disproportionately among women. From 2008 to 2012, the liver cancer-related death rate increased by an average 2.8% per year among men and 3.4% per year among women.

Race/Ethnicity	2020 Number	2020 Rate
White, non-Hispanic	9,397	3.18%
Black, non-Hispanic	2,743	5.63%
Hispanic	1,979	4.00%
Asian/Pacific Islander, non-Hispanic	324	1.44%
American Indian/Alaska Native, non-Hispanic	308	10.17%

Figure 5. 2020 Hepatitis C related deaths by number and rate by race/ethnicity. *rates are age-adjusted per 100,000 US standard population. Source: Center for Disease Control

Although the United States has one of the more robust surveillance and health systems in the world, epidemiological data reveals striking inequities in hepatitis C burden. The lack of response to address the most vulnerable populations is just one roadblock to the eradication of hepatitis C in the United States. In the next part of this series, we will cover another roadblock to eradication—the inaccessibility of treatment for the Hepatitis C Virus.

Direct Acting Antiviral Drugs For The Treatment Of Hepatitis C

⁓

In the previous chapters of this book, we've outlined the seriousness and uncontrolled nature of hepatitis C. In the absence of a vaccine, the control and treatment of hepatitis C, much like that of HIV/AIDS, depends on the use of antiviral medications. Recent progress in the development of several highly active, anti-hepatitis C drugs has been a triumph of modern medicine. These drugs usher in an era of effective treatment and even elimination of hepatitis C both nationally and globally. Later in this series I will describe how these drugs have been used to eliminate hepatitis C from some countries and discuss how that success may be replicated elsewhere.

History

The first effective treatment for hepatitis C infection, alpha interferon, was introduced in 1986. Interferons are naturally occurring proteins in the body that mobilize the body's natural immune system. Interferons do not act directly on the virus itself but rather activate the hosts' antiviral immune defense. This early version of immunotherapy did have some clinical benefit but was largely ineffective against more advanced hepatic C induced liver disease. Treatment with alpha-interferon is accompanied by serious adverse events including, including hair loss, gum disease, and flu-like symptoms.

In the mid 1990s the drug ribavirin was added to the treatment regimen. Whereas interferon acts by activating the immune system, ribavirin interrupts virus replication, presumably by inhibition of the polymerase. The combination of the two drugs resulted in sustained suppression of the infection for up to half of those treated. The drug combination is not equally effective against all hepatitis C strains. The addition of ribavirin boosted the effectiveness of the alpha interferon therapy but also came with an additional risk of serious side effects including thyroid issues, anemia, and psychosis.

The next generation of hepatitis C drugs appeared in the early 2000s in the form of long-acting pegylated alpha interferons. Pegylation is the process of increasing the molecule size of an interferon by adding polyethylene glycol molecules. The larger interferon molecules slow down the rate in which the drug is absorbed, thus increasing the amount of time the drug remains in the body. These long acting pegylated interferons had higher response rates than earlier interferon therapy, especially when combined with ribavirin. They also needed to be injected fewer times than interferon monotherapy which lessened side effects.

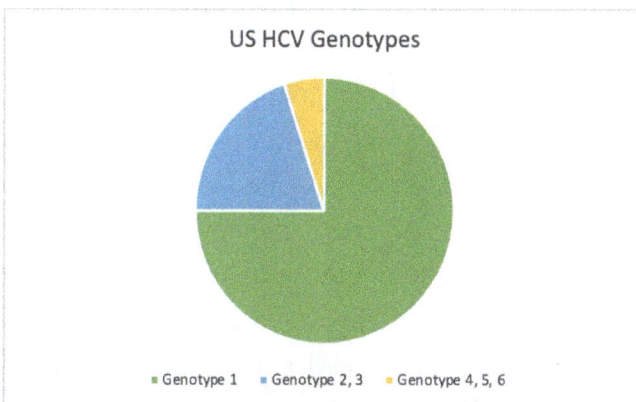

Figure 1. Distribution of HCV genotypes in the United States.

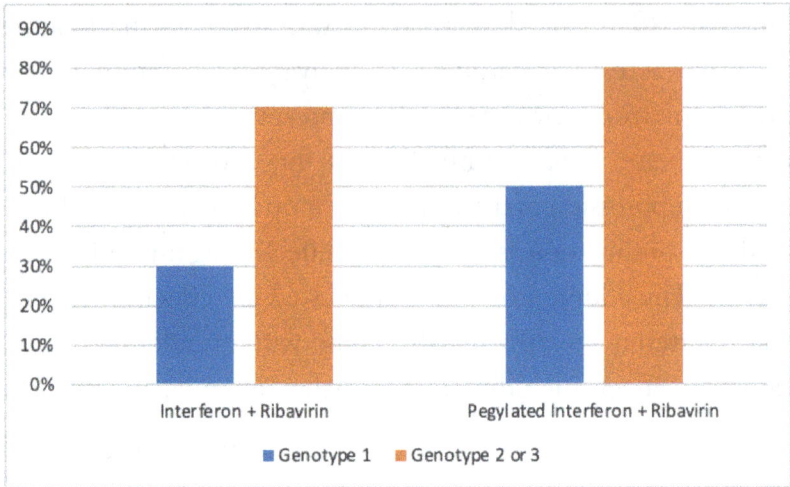

Figure 2. Treatment efficacy of interferon, pegylated interferon, and ribavirin.
Source: Manns et al, 2001

One of the major shortcomings of early hepatitis C drugs was their various effectiveness across different hepatitis C virus subtypes. In the United States, approximately 75% of chronic hepatitis C infections are caused by hepatitis C genotype 1, 15 to 20% by genotype 2 or 3, and less than 5% by genotypes 4, 5, or 6 (Figure 1). The combination therapy of alpha interferon and ribavirin increased the efficacy of interferon monotherapy from around 15% to 40%, but response varied by hepatitis C virus genotype (Figure 2). Only an average of 30% of those with subtype 1 were able to achieve a sustained viral response while 70% of those with subtypes 2 and 3 were able to achieve a sustained viral response. The effectiveness of combination long-acting pegylated alpha interferon and ribavirin also varied across genotypes. Treatment was least effective in those with genotypes 1 and 4, and only around half of patients with those subtypes were able to achieve a sustained viral response. Those with genotype 2 and 3, however, had a sustained viral response rate of around 80% and in half the amount of treatment duration.

Current Drugs

As the need for a more curative hepatitis C treatment grew, drug companies began investing in developing a new type of drug—one that could offer a more sustained viral response across all genotypes. This new generation of drugs, called direct-acting antivirals, were designed to attack specific molecular targets known to play a role in the replication of the hepatitis C virus. In 2011, the first direct-acting antivirals for hepatitis C, telaprevir (developed by Vertex pharmaceuticals and Johnson and Johnson) and boceprevir (developed by Merck), were approved by the Federal Food and Drug Administration. These first generation direct-acting antivirals offered a more sustained virological response, especially among those with genotype 1, as well as a shorter duration of treatment, and fewer adverse effects. Today's direct-acting antivirals are even more effective and are classified by which part of the hepatitis C virus' replication cycle they target (Figure 3). There are 3 classes of direct-acting antivirals: inhibitors of the NS3/4A protease, those that target the NS5A replicase factor, and those that target the NS5B RNA polymerase.

NS3/4A protease inhibitors target the NS3/4A protease enzyme which is needed for the virus to develop the proteins necessary to replicate itself. These drugs are mainly used to treat genotype 1 of the hepatitis C virus but may also be used in combination with other drugs to attack the virus from multiple replication sites. Compared to other classes of direct-acting antivirals, NS3/4As have longer regimens, more side effects, and are more susceptible to the hepatitis C virus developing a resistance. NS3/4A protease inhibitors currently approved to treat hepatitis C include: Grazoprevir, Paritaprevir, Voxileprevir, and Glecaprevir.

NS5A is a multifunctional protein important for the replication of the viral genome. NS5A inhibitors work by blocking the virus's ability to assemble new virions. These inhibitors are effective against all virus genotypes but can also be poorly tolerated and are susceptible to resistance. These drugs are also known to work better when combined with either alpha pegylated interferon or ribavirin. Generic names of NS5A protein inhibitors include: Elbasivr, Ledipasvir, Ombitasvir, Velpatasvir, and Pibrentasvir.

NS5B is a membrane anchored enzyme that initiates the synthesis of the hepatitis C virus's RNA genomes. NS5B inhibitors stop replication of the hepatitis C virus by blocking the function of that enzyme. These drugs work well across all genotypes, are generally well tolerated, and the least susceptible to viral resistance. Current NS5B inhibitors used to treat hepatitis C include: Sofosbuvir and Dasabuvir.

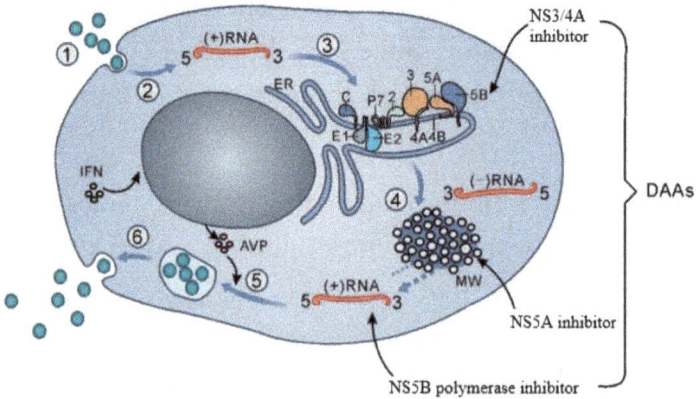

Figure 3. The direct-acting antiviral targets in the hepatitis C virus replication cycle. Source: Hayes et al, 2018

Combinations of Direct-Acting Antivirals

Like interferon therapy, optimal treatment with direct-acting antivirals includes consideration of a patient's genotype and comorbidities such as advanced cirrhosis and coinfection with hepatitis B or HIV. Also, like earlier therapies, many direct acting antivirals are more effective when combined with other antivirals. Much like the human immunodeficiency virus, the hepatitis C virus mutates quickly to develop a resistance to the drugs used to attack it. When used in combination with other antivirals, including ribavirin, direct-acting antivirals can inhibit more than one replication target and increase efficacy. The market for these new combination drugs was also incredibly lucrative. In 2011, Gilead acquired the pharmaceutical company Pharmasset for $11 billion and produced the first hepatitis C combination pill, Harvoni, which was approved in 2014. Harvoni combined two direct-acting antiviral therapies (ledipasvir and sofosbuvir) and was the first FDA-approved regimen that didn't require the addition of pegylated interferon injections and ribavirin. Today, many other combination pills are available and can be prescribed according to hepatitis C virus genotype.

In the absence of a preventative hepatitis C vaccine, direct-acting antivirals have proven to be very effective in treating active hepatitis C virus infections. Timely, appropriate treatment for hepatitis C, however, remains an important part of disease elimination. In the next part of this series, we'll cover the clinical protocol for screening and treating hepatitis C with direct acting antivirals.

A Cure For Hepatitis C: Diagnosis And Direct-Acting Antiviral Drugs

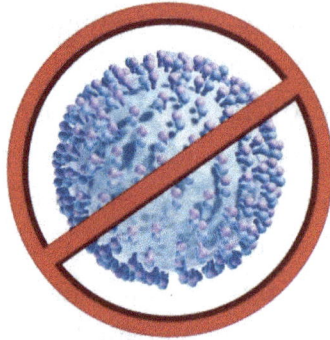

In previous articles of this series, we've discussed several aspects of the hepatitis C virus including the nature of the infection and its distribution both globally and in the United States. Most recently, we've discussed the history of direct acting antiviral drugs for the treatment of hepatitis C. Here we describe the use of those drugs to treat the disease. Remarkably, these drugs can not only cure hepatitis C in individuals but also eliminate the disease in entire populations. In this article, we'll cover the clinical treatment protocols for hepatitis C and its importance in providing optimal care to each patient.

Hepatitis C Diagnostic Tests

Hepatitis C is still endemic to most of the world, but it need not remain so. The disease has been eliminated in several counties, like Egypt, with widespread diagnostic screening and accessible treatment. However, before treatment is initiated, it is critical to properly diagnose and ascertain if treatment is needed at all. Some people with hepatitis C can clear the infection on their own while the majority, if left untreated, can have long term virus replication. These chronic infections can lead to serious liver disease, cancer, and other serious consequences. A streamlined diagnosis process is, therefore, crucial to the United States' eradication efforts. The first step in hepatitis C diagnosis is to screen for exposure to the hepatitis C virus. Exposure is determined by measuring the presence of hepatitis C antibodies in the blood which are produced whenever a person is infected regardless of whether the infection cleared on its own. This is done by performing a rapid test much analogous to the one used to diagnose COVID-19. These rapid assays use a finger prick and can produce accurate results in as little as 20 minutes. A positive result from this test indicates the presence of hepatitis C antibodies and confirms an infection (Figure 1).

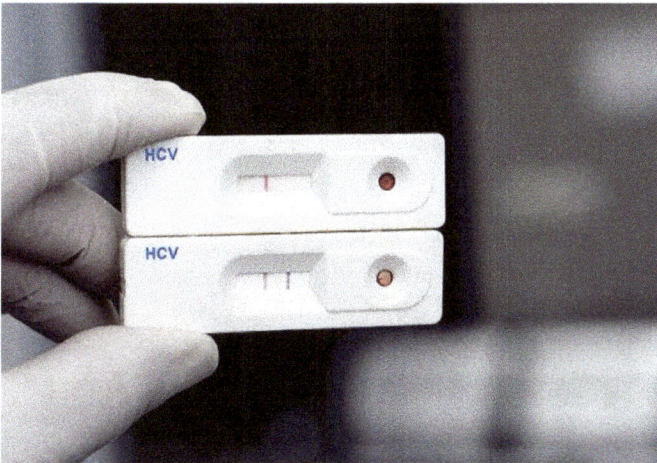

Figure 1. Picture of a hepatitis C antibody rapid test. This test uses a fingerpick to detect antibodies. The above test is a non-reactive negative result, and the bottom test is a reactive positive result.

After confirmation that hepatitis C antibodies are present, an RNA test needs to be performed to determine if there is active virus replication. This test uses a polymerase chain reaction to measure the levels of hepatitis C virus RNA in the blood. Again, this test is analogous to the one used to diagnose COVID-19 and is widely available. These tests also determine the levels of viral RNA in the blood also known as the viral load. A high viral load means that the virus is actively replicating, and the infection is current. A small percent of the population, however, may have hepatitis C antibodies present but a low or undetectable viral load. This is mostly likely because the person has undergone spontaneous clearance, or their body's natural immune system was able to clear the infection on its own. Both the antibody and RNA tests are highly specific and sensitive, making false positives and false negatives rare occurrences. After a confirmed active hepatitis C infection, an additional nucleotide genotyping test should be performed to determine the genotype of the viral infection (Figure 2). Determining both the genotype and viral load will determine the best course of treatment.

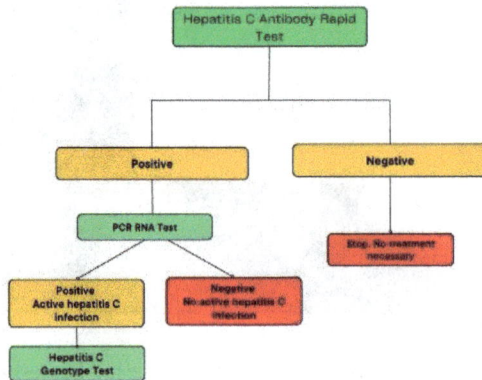

Figure 2. The hepatitis C diagnosis cascade. Diagnosis begins with a rapid assay to detect the presence of hepatitis C virus antibodies. A PCR test is then

performed to determine the levels of viral RNA in the blood. The presence of both hepatitis C antibodies and hepatitis C RNA indicates an active infection. Some cases will show the presence of antibodies but little to no RNA which indicates probable spontaneous clearance.

The Right Drug for the Right Genotype

From previous articles, we should be familiar with the importance of matching the right direct acting antiviral to the right hepatitis C strain. For example, with COVID-19, antibodies only work for some but not all virus variants. Hepatitis C, although not as apparently variable in the population, does come in many different variants. Each variant responds somewhat differently to each of the hepatitis C drugs. For example, several of the strains endemic to Egypt and India are relatively resistant to interferon and ribavirin treatments. More recent data has shown that, after a confirmation of active hepatitis C viral replication, it is critical to determine the specific strain of hepatitis C, called the genotype.

Genotypes are determined using genomic nucleotide sequencing assays. These tests use PCR to amplify and sequence hepatitis C virus RNA to detect sequence variations in the virus's Core and NS5B protein regions. There are currently 6 different hepatitis C genotypes present across the globe. In the United States, genotypes 1, 2, and 3 make up over 98% of infections . Some genotypes can be further divided into subtypes. Genotype 1, for example, has subtypes 1a and 1b. Genotype can also affect disease progression in addition to treatment response. Those with subtype 1b may be at higher risk for developing cirrhosis, and those with genotypes 1b and 3 have higher chances of developing liver cancer. It is possible to be coinfected with more than one hepatitis C genotype or subtype which is called a mixed infection. Mixed infections may require more robust sequencing methods and more tailored treatment regimens. Recommended treatment regimens by subtype can be seen in Figure 3. As drug development progresses, many newer direct-acting antivirals are designed to be pangeotypic and can treat all genotypes with equal success.

Brand Name	Generic Name	Class	Genotype
Harvoni	• Ledipasvir • Sofosbuvir	• NS5A polymerase inhibitor • NS5B nucleotide polymerase inhibitor	1, 4, 5, 6
Epclusa	• Sofosbuvir • Velpatasvir	• NS5A polymerase inhibitor • NS5B nucleotide polymerase inhibitor	1-6
Vosevi	• Sofosbuvir • Velpatasvir • Voxilaprevir	• NS5A polymerase inhibitor • NS5B nucleotide polymerase inhibitor • NS3/4A protease inhibitor	1-6
Sovaldi	• Sofosbuvir	• NS5B nucleotide polymerase inhibitor	1-4
Zepatier	• Elbasvir • Grazoprevir	• NS5A polymerase inhibitor • NS3/4A protease inhibitor	1,4

Figure 3. Table of direct acting antiviral regimens and the genotypes they are best suited to treat. Many regimens are geared toward treating genotypes 1, 2, and 3 which are the most common in the United States.

Treatment Regimens and Outcomes

Most direct-acting antiviral regimens consist of taking a daily pill for an average of 3 months with interval RNA testing performed to assess how well the treatment is working. The end goal of treatment is to achieve a sustained viral response. This is measured by an undetectable hepatitis C virus RNA level, typically under 25 IU/mL, at least 3 months after completing therapeutic treatment (Figure 4). Achieving an undetectable viral load within this timeframe is considered a clinical cure, and 99% of patients who finish a course of direct-acting antiviral therapy achieve a sustained viral response. About 1-2% of patients who are treated with direct-acting antivirals may not achieve a sustained viral response. In those cases, another drug combination may be prescribed and pegylated interferon or ribavirin may be added.

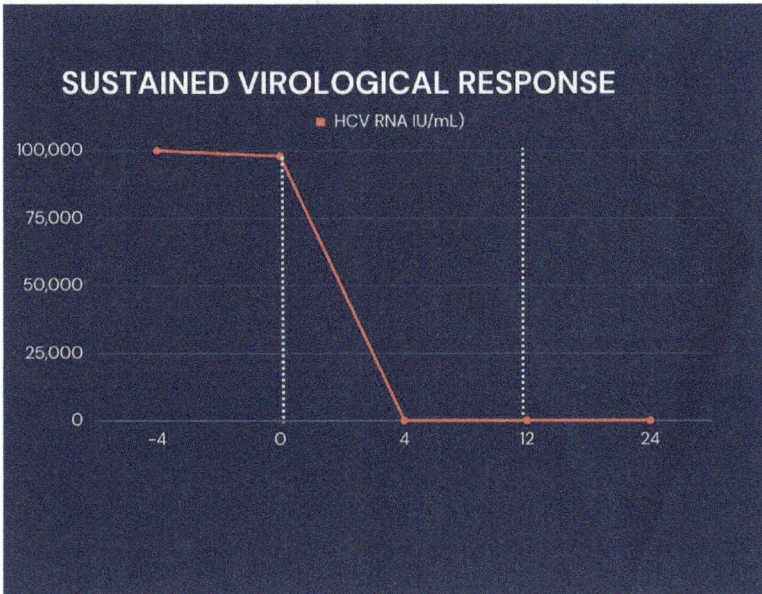

Figure 4. A time graph of direct acting antiviral treatment. Viral RNA levels are measured at the beginning of treatment (Week 0) and after a full course of treatment that lasts around 12 weeks. If viral RNA levels are still undetectable 12 weeks after the completion of treatment, that is considered a sustained viral response and a cured case of hepatitis C.

In addition to virus genotype, the screening process for hepatitis C also involves assessing additional risk factors for each patient, especially when retreating a patient who has been previously treated with a direct-acting antiviral-based therapy. In the United States, the current recommendation for starting direct-acting antiviral treatment is to begin treatment within 6 weeks of a confirmed diagnosis. High- risk factors like co-infections with hepatitis B or the human immunodeficiency virus, significant cirrhosis, or the necessity for liver transplantation indicate that immediate treatment is needed. Unfortunately, in some places, treatment with direct-acting antivirals is both expensive and difficult to obtain. These additional calculated risk factors can affect a patient's likelihood of receiving timely treatment as well as if and which drugs are covered by medical insurance providers.

In conclusion, in the past few decades, we've witnessed something close to a medical miracle in the treatment of hepatitis C. The number of drugs and diagnostic tests developed are well studied and highly effective in producing a permanent cure for each individual patient. As we'll see in the next part of this series, these drugs can also eliminate hepatitis C in whole populations around the world.

How Did Egypt Eliminate Hepatitis C In Less Than One Year?

فيروس "سى" ... ياحنا ... ياهو

Translation: "It is either us, or hepatitis virus C: (1) Don't use personal items that are not yours; (2) It is ok to have contact with HCV patients; (3) Pills are safer than injections; (4) Consult with a physician and test for HCV; and (5) HCV treatment is available and effective." Source: Hassanin et al, 2021

Both the World Bank and the World Health Organization declare Egypt to have eliminated hepatitis C from its entire population. How was this done, and why hasn't it been done in the United States or any other country regardless of income? Our analysis of the Egyptian program suggests there are three main barriers to eliminating hepatitis C. Most important of these is political will. The second is the organization of public health systems to implement the program and finally, the third barrier is the cost of the diagnostic tests and treatment. Here we provide some details of

Egypt's 100 Million Healthy Lives campaign which, with a modest loan from the World Bank, resulted in the elimination of hepatitis C.

How Did Egypt Become the Country with the Highest Incidence of Hepatitis C?

Many people in Egypt were infected with hepatitis C as children. During a mass vaccination program against schistosomiasis, children were inoculated with anti-schistosomiasis drugs using cross contaminated needles. After nearly thirty years of the campaign, Egypt realized that sharing used needles was rapidly transmitting the hepatitis C virus and ended the campaign. By that time, almost 15% of the population had been infected. The long term consequences of this were a high incidence of serious liver disease and one of the highest incidences of liver cancer in the world.

In response to the high prevalence of hepatitis C, the Egyptian government designed the 100 Million Healthy Lives program. The campaign was enabled by a loan of approximately $250 million which was around half of a $530 million loan from the World Bank. The seven month program was formed under the leadership of President Abdel Fattah al Sisi, managed by Egypt's Ministry of Health, and the detailed plans were designed and implemented by Dr. Wahid Doss. Its ambitious objective was to screen and treat Egypt's entire population over the age of 12, including those in prison and in the army, for hepatitis C.

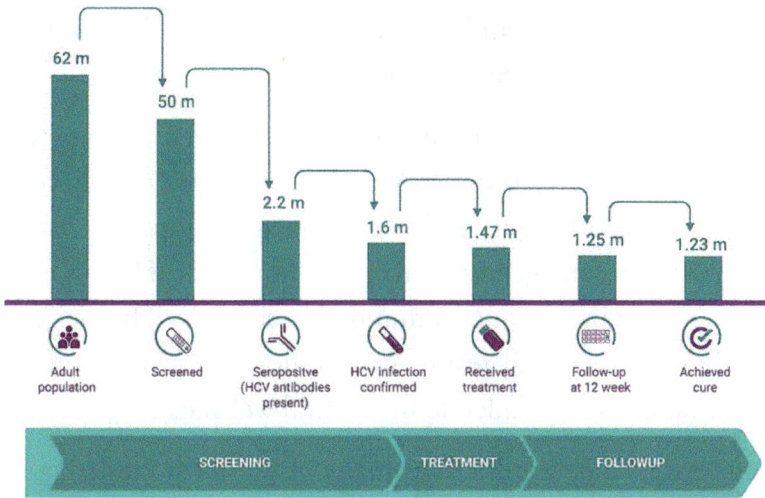

Figure 1. Egypt's flow of individuals from screening to treatment to follow up during the 100 Million Healthy Lives Campaign. Source: Hassanin et al, 2021

Design of the Program

The 100 Million Healthy Lives campaign was a universal program established throughout the country and illustrated the president's commitment to eliminating hepatitis C. Egypt's ability to organize its efforts across the country made the 100 Million Healthy Lives program the largest medical screening campaign to date. The program followed a simple, yet highly effective cascade of 6 steps. (Figure 1).

The first step was to organize the screening groups. For the first phase of the program, the government targeted citizens 18 years and older but later expanded the program to include those over the age of 12.

The second step was to recruit people for screening. The Egyptian government employed several promotional avenues, including a mass media campaign and print media, to spread the word about the program and encourage people to get tested.

The third step was to screen the 50 million citizens who opted for hepatitis C testing for seropositivity. Citizens were able to register for appointments on the centralized campaign website, www.stophcv.eg, and get tested at one of thousands of testing centers and mobile units stationed around the country. The fingerstick antibody tests produced results in as little as twenty minutes and were immediately uploaded to the central website.

Step four was to determine how many of those who tested positive for hepatitis C antibodies had active infections. Those with positive antibody results were referred to specialized testing centers for confirmation PCR testing. The Egyptian government established several PCR testing centers throughout the country, some of which were able to conduct an estimated 36,000 tests each day.

Step five was to provide those with active infections with three months of free treatment. On average, approval for treatment took place within a week of PCR confirmation and was dispensed on site at specialized hepatitis C centers. Of the 1.6 million adults with active hepatitis C infections, 1.47 million accepted government provided treatment for no cost, or opted for private treatment for $70.

The final step was to retest those who received treatment for clearance of the disease. After three months of treatment, individuals were reevaluated for disease clearance with an additional PCR test. Those who tested negative were then provided with a certificate of cure. At the end of the campaign, an estimated 1.23 million Egyptian citizens with active hepatitis C infections were cured.

Figure 2. A 100 Million Healthy Lives advertisement banner featuring President El-Sisi. Source: Heliopolis University

Figure 3. A 100 million healthy lives mobile testing van. Source: Egypt Today

Overcoming the Barriers

One of the major contributors to the success of Egypt's program was the country's political will to eliminate hepatitis C. With the strong support of both President El-Sisi and the Health Ministry, Egypt's dedication to eliminating hepatitis C made the country an ideal recipient for a collaborative partnership between the Egyptian government and the World Bank. To illustrate the country's commitment to elimination, the government made a great effort to ensure everyone knew the program had the full support of both the President El-Sisi and the Ministry of Health. The country was flooded with advertisements for the campaign. Banners and billboards were installed everywhere from the smallest to the largest cities (Figure 2). In addition to word of mouth and print promotion, the government also used television commercials, SMS text messaging, and social media platforms like Facebook to advertise the campaign. The demonstrated will of Egypt's government encouraged millions of its citizens to participate in the program and made the 100 Million Healthy Lives campaign one of the largest health screening programs in the world.

The second barrier most countries face in eliminating hepatitis C is the need for coordinated organization. The Million Healthy Lives program was made possible by creating a heavily resourced flow of individuals from screening to treatment. In addition to establishing over 18,000 testing centers throughout the country, the government also created over 1,000 mobile testing vehicles and trained over 60,000 enthusiastic staff. Each testing site consisted of a young physician, a nurse, a data specialist, and several newly trained members of local communities. (Figure 3). The amount of personnel and wide distribution of hepatitis C centers enabled them to reach all its targeted 50 million citizens, including the 57% that lived in rural areas. The government also used a virtual private network within its national healthcare service to report all the results of the campaign in real time. Program officials were able to track how many people were screened, what the age distribution was, and the percent of the population covered. The central organization system also allowed for the screening of other public health diseases in Egypt including hypertension, diabetes, and obesity. Egypt's

organizational efforts provide an excellent model for screening and treating public health issues across entire populations.

For most countries, the cost barrier to eliminating hepatitis C is substantial. The success of Egypt's program relied on the ability of its government to control the costs of implementation, diagnostics, and drugs. To control the organizational costs, Egypt was able to deploy much of its existing infrastructure within its centralized healthcare service system. The government was also able to establish testing centers, train its staff, and establish its specialized hepatitis C website all for a low cost of about 1000 Egyptian pounds per day.

Egypt's program included several points of testing including the seropositivity rapid test, the active infection PCR test, and the secondary PCR test to verify disease clearance. While those tests are 10 to 100 times more expensive in the United States, the Egyptian government was able to bring the costs of diagnostics down to a considerably lower price. During the program, Egypt was able to provide antibody tests for 56 cents and PCR tests for just $5 USD. These same tests in the United States can cost around $20 for an antibody rapid test, and up to $300 for a PCR test.

Internationally, many hepatitis C drugs are available at lower costs through pharmaceutical tier systems. Direct acting antivirals are available for between $22,000 to $87,000 for high income countries, $6000 for middle income countries, and $900 for low income countries. For the success of Egypt's program, however, the Egyptian government needed an even greater reduction in cost. The Egyptian government declared the burden of hepatitis c drug costs as a contributor to its endemic status. This exempted the country from drug patent protection in accordance with the Trade-Related Aspects of Intellectual Property Rights agreement. Egypt was then able to purchase active pharmaceutical ingredients for seven different direct acting antivirals from India where treatment can cost anywhere from $25 to $35 USD. These materials were then formulated by Egyptian pharmaceutical companies. This allowed the program to provide free treatment for all its citizens at a cost of $45 to the Egyptian government. These efforts also allowed those who opted for private treatment to pay just $70.

For comparison, those same treatment regimens can cost a United States citizen anywhere from $24,000 to $84,000. In total, the testing and treatment portion of the 100 Million Healthy Lives program cost the Egyptian government around $207 million USD or about $184 USD per person.

What Made the 100 Million Healthy Lives Program So Effective?

According to the World Health Organization's verification report, the success of the 100 Million Healthy Lives campaign was achieved through six key factors. The first was its establishment of a specialized committee to support national public health initiatives. In 2018, the Ministry of Health formed such a committee for the sole purpose of designing and overseeing the program and awarding proper allocation of resources.

The second was its commitment to updating national health strategies based on its national priorities. In conjunction with the World Health Organization and the World Bank, Egypt's government developed a plan of action for eliminating hepatitis C and effectively outlined its objectives and methods to achieve those goals. Third, Egypt reviewed and updated its national hepatitis C relevant legislation and regulations which enabled the government to sustain its achievements.

Another key factor was the program's ability to finance and provide resources for sustainability. Egypt's ability to mobilize its human resources by conducting widespread training and developing standard operating procedures for screening and treatment enabled the program to equally serve citizens across all regions of the country. Additionally, by ensuring the existence of a national medication use and procurement policy, Egypt was able to dramatically reduce the costs of treatment drugs and now has the largest share of the pharmaceutical industry in the Middle East and North Africa region. Lastly, the government created mechanisms like its patient input hotline which allowed proper, prompt, and continuous adjustment and improvement of the program.

The specifications of the Egyptian program are no secret and have been described in detail by medical and scientific literature including the New England Journal of Medicine. Its success was also recently verified by the World Health Organization in their report detailing its creation, implementation, and success. In any country, Egypt's 100 Million Healthy Lives program has the feasibility to eliminate hepatitis C as well as identify other serious health issues including hypertension, diabetes, and obesity. The remaining question is why aren't similar programs being established in other high, middle, and even low income countries?

Why Hasn't The United States Eliminated Hepatitis C When Egypt Did So? Why Are The United States' Plans To Eliminate Hepatitis C 10 Times More Expensive Per Person And 10 Times Longer?

	2010	2023
Egypt	5.5 Million	Less than detectable
United States	3.2 Million	2.5 Million

Figure 1. A comparison of the number of viremic hepatitis C infections in the United States and Egypt in 2010 and 2020. Since the conclusion of the 100 Million Healthy Lives Campaign, Egypt has reached the WHO hepatitis C targets for the elimination. Source: Access Health

Three years ago, Egypt and the United States had approximately the same number of people infected with chronic hepatitis C. The hepatitis C burden in Egypt was also three to four times higher per capita than in the United States. Today, the situation is very different. Egypt has been certified by the World Health Organization to be virtually free of the disease. The United States, however, still carries the same burden despite having the resources to eliminate hepatitis C (Figure 1). In the previous story, we saw how Egypt's 100 Million Healthy Lives program achieved country wide elimination by demonstrating the will and organization to do so, as well as finding a way to cover the costs. Unfortunately, the

United States seems to lack those same tenets. Recently, the current administration took a major step forward and laid out a historic $11 billion plan to eliminate hepatitis C in the United States within five years.

The $11 Billion Dollar Plan

The barriers that the United States faces to eliminate hepatitis C are comparable to the barriers that Egypt faced. The United States must also demonstrate a strong will to eliminate the disease, organize its infrastructure and resources, and find ways to lower the high costs of diagnostics and drugs. The success of Egypt's 100 Million Healthy Lives campaign provided an excellent model for creating a hepatitis elimination plan, and in March of this year, the current administration revealed its Nation Hepatitis Elimination Program. Outlined in the government budget for fiscal year 2024, this five year and $11.3 billion plan is a declaration of the United States' willingness to eliminate hepatitis C. It establishes a cohesive effort that brings together all facets of the healthcare delivery system including state governments, federal agencies, and community health providers.

The National Hepatitis Elimination Program plan outlines four main objectives that aim to "increase access to curative medications and expand implementation of complementary efforts such as screening, testing, and provider capacity." These objectives are designed to tackle all three of the main elimination barriers: the will, the organization, and the cost.

Unlike Egypt, the United States doesn't have a national health service and as a result, there isn't a standardized protocol for surveilling, identifying, and treating hepatitis C. Even with the

CDC's universal testing guidelines, 40% of the United States' population is unaware of their infection status until severe, often irreversible liver damage occurs. This is partially due to the lack of coordination among states. While some states may possess the will and the resources necessary to inform their populations about hepatitis C testing and treatment, others might not demonstrate the same urgency or have the resources to reach its entire population. The first tenet of this program addresses this by prioritizing finding ways to reach and engage people, especially those in marginalized communities. In addition to nationwide media campaigns, telehealth is a promising outreach tool. It allows providers to remotely reach individuals who would otherwise not have access to information about preventive measures and the importance of getting tested.

The next step of the program is to test the population. Currently, the diagnostic process for hepatitis C includes both an antibody test and a PCR test that can require multiple provider visits over several weeks to get a diagnosis. This can be a deterrent for individuals to get tested. One way the National Hepatitis Elimination Program plans overcome this barrier is by increasing the availability of point of care diagnostic tests. As we've seen with COVID-19, at home or local point of care testing is a highly effective way to quickly screen a mass number of people and identify individuals who need treatment. Europe, in fact, already has such a test which can readily be employed in the United States. Much like the call to action we observed with COVID, making information, and testing widely accessible is critical to the success of a nationwide elimination campaign.

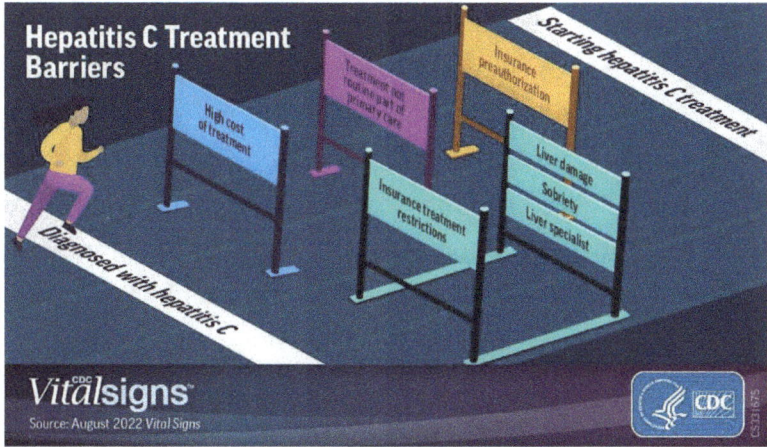

Figure 2. Many insurance providers are unwilling to cover the costs of hepatitis c treatment. Some providers require proof of sobriety and extensive liver damage before covering the cost of direct acting antivirals. Source: Center for Disease Control

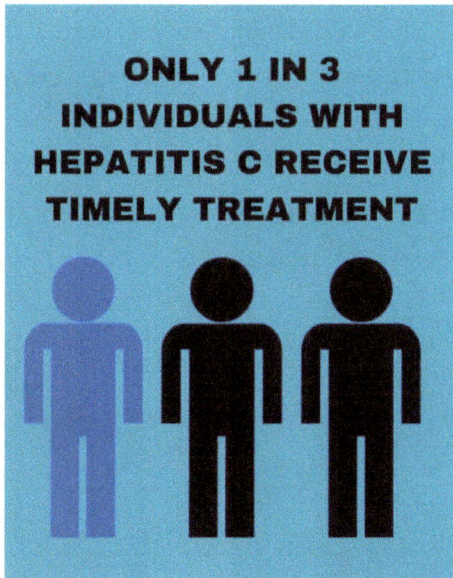

Figure 3. As a result of low insurance coverage rates, only 1 in 3 individuals with hepatitis C receive timely treatment

The curative ability of direct acting antivirals is a modern medical miracle, but the high costs of the drugs make it widely unavailable, especially for those who are incarcerated or on Medicaid. Due to the high financial barrier, many health insurance providers require proof of liver damage or sobriety to cover the cost of treatment with direct acting antivirals. As a result, only about a third of individuals with active hepatitis C infections receive treatment within one year of diagnosis (Figures 2 and 3). This not only prevents many marginalized individuals access to life saving treatment, but also prevents pharmaceutical companies from making money off drugs that very few can afford to use. The National Hepatitis Elimination program proposes a subscription model solution, also known as the "Netflix" model, that benefits both parties (Figure 4). In a subscription model, states pay a pharmaceutical company a lump sum for free access to their hepatitis C drugs for a specific population. This model was piloted in Louisiana in 2019 and regarded as a success as the Louisiana state government was able to drastically expand access to hepatitis C treatment for its Medicaid and incarcerated populations. If deployed on a national scale, the subscription model can help insurers lower coverage costs and provide life saving care to all that need it.

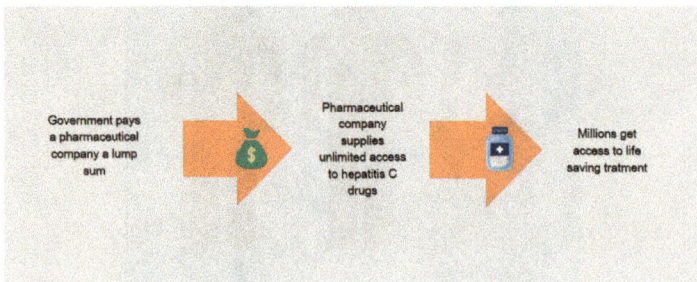

Government pays a pharmaceutical company a lump sum → Pharmaceutical company supplies unlimited access to hepatitis C drugs → Millions get access to life saving tratment

Figure 4. The subscription, or "netflix" model allows state health programs to lower the cost of hepatitis c treatment by providing pharmaceutical companies with an upfront lump sum in exchange for unlimited access to their drugs.

The final pillar of the National Hepatitis Elimination plan is to reignite research for a hepatitis C vaccine. While direct acting antivirals are highly effective at curing hepatitis C, they do not prevent reinfections. In conjunction with a national awareness and prevention campaign, however, a hepatitis C vaccine would reduce the rate of both new and recurrent infections. This key step not only addresses the current hepatitis C endemic but can also prevent future waves of infection.

Through the four major pillars of the program, the National Hepatitis Elimination Program addresses the will, organization, and cost barriers that hinder elimination of hepatitis C in the United States. The plan outlines a cohesive effort among many stakeholders to create a cascade of identifying and curing those with hepatitis C—an endeavor that will save billions in future healthcare costs. So why does the program need five years and $11 billion? The United States' program must reach nearly three times the population as Egypt and has many start-up costs including lump sums for pharmaceutical companies, a nationwide awareness campaign, vaccine research, and labor costs. Although the scale of the National Hepatitis C Elimination program may seem daunting, it is worth it. We know it is possible to eliminate hepatitis C in the United States, and now that there's a plan, we just need to put it into motion.

It is also worth noting that the 100 Million Healthy Lives campaign not only screened the entire population over 12 for hepatitis C, but also for diabetes, hypertension, and obesity. Egypt's national public health service also offered free treatment for diabetes and hypertension and free counseling for obesity. We can only hope that the United States achieves similar efficiency in disease screening such as screening women for breast and cervical cancer.

Is A Broadly Protective And Long Lasting Hepatitis C Vaccine Possible?

~~~~~~

Hepatitis C is a serious blood borne disease that can cause severe liver damage and impacts over 2.5 million Americans. While the disease can be treated, treatment is not yet universally available and affordable to those who need it. With an estimated 66,700 new infections occurring in the United States each year, a hepatitis C vaccine would help to eliminate the disease from the United States and elsewhere. Current understanding of the natural history and greater variability of the hepatitis C virus means that the vaccine needs to be both broadly protective and long lasting. Is such a vaccine feasible? Here we outline the reasons for optimism despite the prevalence of hepatitis C variants which have confounded previous vaccine attempts.

Hepatitis C Variants: The Major Difficulty in Creating a Vaccine

The hepatitis C virus replicates rapidly and generates high variability resulting in multiple strains and subtypes (Figure 1). There are currently eight globally recognized hepatitis C genotypes with an estimated 30% variance of nucleotide divergence between them. Each genotype also carries multiple subtypes that can have up to 20% variance between them. For comparison, the hepatitis B virus, for which we do have a vaccine, has only an estimated 8% variance between genotypes. It is also important to note that the hepatitis C virus can cause superinfections. In these cases, an individual can be infected with multiple types or subtypes of the virus at the same time. Evidence suggests that infection with one strain does not protect against infections of another strain. The implication of this is that a vaccine would need to be broadly neutralizing against many different strains at the same time.

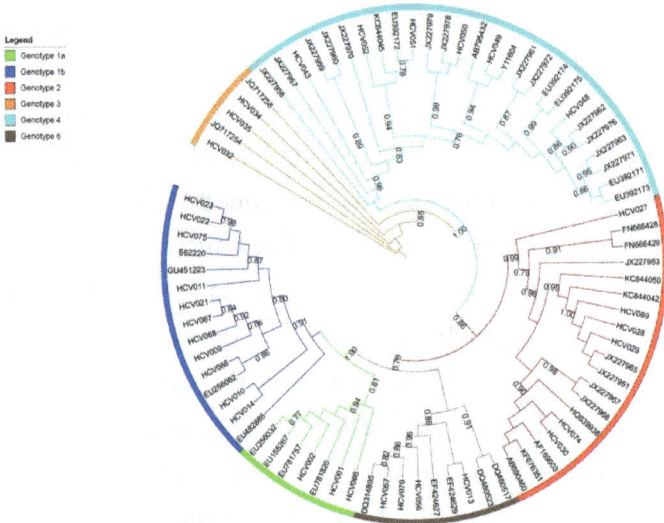

Figure 1. A phylogenetic tree of the NS5B gene sequences of 6 of the major hepatitis C genotypes. Source: Goletti et al, 2019. License: Attribution 4.0 International.

Reasons for Optimism

Despite the presenting challenges, there are reasons for optimism in creating a vaccine for hepatitis C. Around 30% of people who are infected with the hepatitis C virus do mount an effective immune response that protects them from disease and clears the infection on its own. The study of these people reveals two major factors about their immune responses: they produce broadly neutralizing antibodies, and they make robust and sustained T-cell responses. The knowledge of how these people can control and clear the infection offers valuable insight for creating a successful vaccine.

Considerations for a Broadly Neutralizing Vaccine

An effective hepatitis C vaccine must first generate broadly neutralizing antibodies and second, elicit long term memory T-cells. In the antibody immune response, B-cells can induce the production of broadly neutralizing antibodies. These specialized antibodies are useful because they can recognize and block viral entry across multiple strains of hepatitis C. In addition to those antibodies, it is also important that the vaccine initiate robust T-cell immune responses. When an adaptive cell-mediated response starts taking effect, the immune system releases two types of T-cells: helper CD4+ cells, which prompt B-cells to make antibodies, and cytotoxic CD8+ cells which kill cells already infected with the hepatitis C virus. Sustained memory T-cells are an important part of clearing current hepatitis C infections and preventing reinfections. Fortunately, we can take lessons from earlier vaccines like the ones we have for the respiratory syncytial virus and the Ebola virus that show that if you understand pre-fusion complexes and conserved epitopes, you can make broadly neutralizing antibodies and strong T-cell responses.

Conserved T-Cell Epitopes

One method that vaccines, like the Ebola vaccine, use to induce production of T-cells is by targeting conserved epitopes. An epitope is a specific part on the surface of a virus that T-cells can bind to and signal B-cells to produce hepatitis C antigen specific antibodies. Conserved epitopes are T-cell targets that appear across multiple strains of hepatitis C which would allow the vaccine to work against a breadth of antigen variants (Figure 2). Studying the antigens of individuals who have spontaneously cleared the hepatitis C virus can offer hints as to which conserved epitopes might be the most useful in inducing the best T-cell responses. This method was also integral to developing a pan-genotypic COVID-19 vaccine.

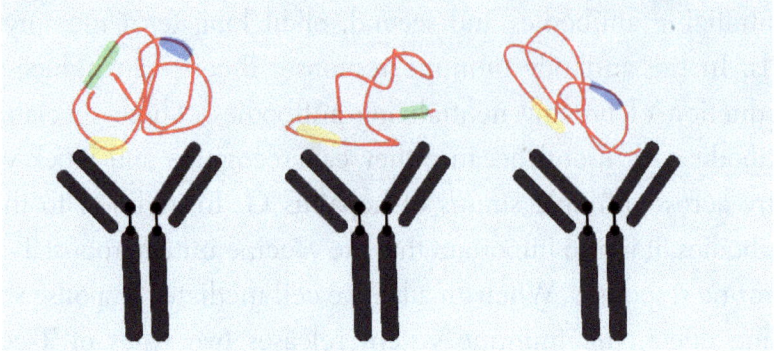

Figure 2. Conserved epitopes are regions of antibody targets that can be found on multiple antigens (yellow highlight). Finding the conserved epitopes of the hepatitis C virus will help develop a broadly effective vaccine that induces T-cell responses against multiple genotypes.

Pre-Fusion Complexes

The success of the respiratory syncytial virus vaccine offers insight into pre-fusion complexes and how they generate broadly neutralizing antibodies. To fuse with and infect liver cells, the surface proteins of the hepatitis C virus must undergo a conformational change. Broadly neutralizing antibodies prevent viral infection by binding to epitopes on virus' surface proteins and blocking the conformational shift from occurring. Locking the virus in its pre-fusion state can then prompt the production of more broadly neutralizing antibodies. A vaccine that contains these pre-fusion antigens would produce large amounts of broadly neutralizing antibodies that not only recognize conserved epitopes, but also bind across multiple targets on the virus' surface proteins. Current research focuses on the AR4 epitope and how it may lock the E1-E2 hepatitis C protein in its pre-fusion form to block the ability to fuse with liver cells.

What Methods of Vaccine Delivery Are Being Considered?

The final consideration for creating a successful hepatitis C vaccine is the design of the vector. There are several methods currently being explored, some of which are more suited for antibody immunity while others are better suited for triggering a cell mediated immune response. One option for delivery is using mRNA as a vector. These types of vaccines use a lipid coating or inactive virus shell to transfer viral RNA into cells to teach them how to make effective antibodies. Another approach relies on subunit vaccines, purified proteins or peptides combined with potent adjuvants. Still others depend on viral vectors such as the family of adeno-associated viruses to deliver the antigens.

While we don't know if these methods will be successful, the current efforts underway are promising. In the meantime, we must rely on the implementation of programs like the 100 Million Healthy Lives initiative in Egypt and the expansion of efforts to lower the costs of hepatitis C drugs.

Coda

If global elimination efforts follow all seven steps mentioned in this book, they will be successful. This book is not only a roadmap for success but it should also motivate our governments and international institutions to act on these issues. We've seen the success of these steps in Egypt and know they can be applied to countless diseases anywhere else in the world. In many cases, however, one or more of these steps are missing. In our next book, we'll look at elimination efforts for other endemic diseases including HIV, tuberculosis, and malaria and discuss which of the seven elements are missing from these programs and how these seven steps can be better applied.

Acknowledgments

William Haseltine, PhD

I would like to thank Kaelyn Varner for her tireless work and good cheer. I could not have attempted this book without the dedicated assistance of the ACCESS Health New York team: Courtney Biggs, Griffin McCombs, Kim Hazel, Amara Thomas, and Roberto Patarca.

This work is supported by ACCESS Health International (www.accessh.org).

Kaelyn Varner, MPH

I would like to acknowledge and express my thanks to my family and friends for their continuous love and support. I would also like to thank Dr. Haseltine for being a tremendous mentor in both publishing this book and in our quest to improve global health.

References

2019 Hepatitis C | Viral Hepatitis Surveillance Report | CDC. (2021, May). 2019 Hepatitis C | Viral Hepatitis Surveillance Report | CDC. https://www.cdc.gov/hepatitis/statistics/2019surveillance/index.htm

2020 Hepatitis C | Viral Hepatitis Surveillance Report | CDC. (2022, August). 2020 Hepatitis C | Viral Hepatitis Surveillance Report | CDC. https://www.cdc.gov/hepatitis/statistics/2020surveillance/hepatitis-c.htm

2021 Hepatitis C | Viral Hepatitis Surveillance Report | CDC. (2023, August). 2021 Hepatitis C | Viral Hepatitis Surveillance Report | CDC. https://www.cdc.gov/hepatitis/statistics/2021surveillance/index.htm

Abeyasinghe, R. R., Galappaththy, G. N., Smith Gueye, C., Kahn, J. G., & Feachem, R. G. (2012). Malaria control and elimination in Sri Lanka: documenting progress and success factors in a conflict setting. PloS one, 7(8), e43162. https://doi.org/10.1371/journal.pone.0043162

Aisyah, D. N., Shallcross, L., Hully, A. J., O'Brien, A., & Hayward, A. (2018). Assessing hepatitis C spontaneous clearance and understanding associated factors-A systematic review and meta-analysis. Journal of viral hepatitis, 25(6), 680–698. https://doi.org/10.1111/jvh.12866

Auty, S. G., Griffith, K. N., Shafer, P. R., Gee, R. E., & Conti, R. M. (2022). Improving Access to High-Value, High-Cost Medicines: The Use of Subscription Models to Treat Hepatitis C Using Direct-Acting Antivirals in the United States. Journal of health politics, policy and law, 47(6), 691–708. https://doi.org/10.1215/03616878-10041121

Bailey, J. R., Barnes, E., & Cox, A. L. (2019). Approaches, Progress, and Challenges to Hepatitis C Vaccine Development. Gastroenterology, 156(2), 418–430. https://doi.org/10.1053/j.gastro.2018.08.060

Barber, M. J., Gotham, D., Khwairakpam, G., & Hill, A. (2020). Price of a hepatitis C cure: Cost of production and current prices for direct-acting antivirals in 50 countries. Journal of virus eradication, 6(3), 100001. https://doi.org/10.1016/j.jve.2020.06.001

Brunner, N., & Bruggmann, P.. (2021). Trends of the Global Hepatitis C Disease Burden: Strategies to Achieve Elimination. 54(4), 251–258. https://doi.org/10.3961/jpmph.21.151

Cartwright, E. J., Patel, P., Kamili, S., & Wester, C. (2023). Updated Operational Guidance for Implementing CDC's Recommendations on Testing for Hepatitis C Virus Infection. MMWR. Morbidity and mortality weekly report, 72(28), 766–768. https://doi.org/10.15585/mmwr.mm7228a2

Chhatwal J, Aaron A, Zhong H, et al. Projected Health Benefits and Health Care Savings form the United States National Hepatitis C Elimination Initiative. National Bureau of Economic Research. 2023. doi:10.3386/w31139.

Chou, R., Dana, T., Fu, R., Zakher, B., Wagner, J., Ramirez, S., Grusing, S., & Jou, J. H. (2020). Screening for Hepatitis C Virus Infection in Adolescents and Adults: Updated Evidence Report and Systematic Review for the US Preventive Services Task Force. JAMA, 10.1001/jama.2019.20788. Advance online publication. https://doi.org/10.1001/jama.2019.20788

Committee on a National Strategy for the Elimination of Hepatitis B and C; Board on Population Health and Public Health Practice; Health and Medicine Division; National Academies of Sciences, Engineering, and Medicine; Buckley GJ, Strom BL, editors. Eliminating the Public Health Problem of Hepatitis B and C in the United States: Phase One Report. Washington (DC): National Academies Press (US); 2016 Jun 1. 3, The Elimination of Hepatitis C. Available from: https://www.ncbi.nlm.nih.gov/books/NBK368067/

Cook, N., Turse, E. P., Garcia, A. S., Hardigan, P., & Amofah, S. A. (2016). Hepatitis C Virus Infection Screening Within Community Health Centers. The Journal of the American Osteopathic Association, 116(1), 6–11. https://doi.org/10.7556/jaoa.2016.001

Echeverría, N., Moratorio, G., Cristina, J., & Moreno, P. (2015). Hepatitis C virus genetic variability and evolution. World journal of hepatology, 7(6), 831–845. https://doi.org/10.4254/wjh.v7.i6.831

Fields, B. N., Knipe, D. M., & Walker, C. (2013). Hepatitis C Virus. In Fields Virology: Emerging Viruses (7th ed., Vol. 1, pp. 306–318). essay, Wolters Kluwer, Lippincott Williams & Wilkins.

Flanigan, C. A., Leung, S. J., Rowe, K. A., Levey, W. K., King, A., Sommer, J. N., Morne, J. E., & Zucker, H. A. (2017). Evaluation of the Impact of Mandating Health Care Providers to Offer Hepatitis C Virus Screening to All Persons Born During 1945-1965 - New York, 2014. MMWR. Morbidity and mortality weekly report, 66(38), 1023–1026. https://doi.org/10.15585/mmwr.mm6638a3

Garg, S., Brooks, J. T., Luo, Q., & Skarbinski, J. (2014). 1588: Prevalence of and Factors Associated with Hepatitis C Virus Testing and Infection Among HIV-infected Adults Receiving Medical Care in the United States. Open Forum Infectious Diseases, 1(Suppl 1), S423. https://doi.org/10.1093/ofid/ofu052.1134

Global hepatitis report. (2017, August 9). Global hepatitis report. https://www.who.int/publications-detail-redirect/9789241565455

Goletti, S., Zuyten, S., Goeminne, L., Verhofstede, C., Rodriguez-Villalobos, H., Bodeus, M., Stärkel, P., Horsmans, Y., & Kabamba-Mukadi, B. (2019). Comparison of Sanger sequencing for hepatitis C virus genotyping with a commercial line probe assay in a tertiary hospital. BMC infectious diseases, 19(1), 738. https://doi.org/10.1186/s12879-019-4386-4

Gray, R., McManus, H., King, J., Petoumenos, K., Grulich, A., Guy, R., McGregor, S. (2023, July 23–26). Australia's progress towards ending HIV as a public health threat: trends in epidemiological metrics over 2004-2021 [Conference presentation abstract]. Twelfth International AIDS Society, Brisbane, Australia.

https://programme.ias2023.org/Abstract/Abstract/?abstractid=35
38

Hassanin, A., Kamel, S., Waked, I., & Fort, M. (2021). Egypt's
Ambitious Strategy to Eliminate Hepatitis C Virus: A Case
Study. Global health, science and practice, 9(1), 187–200.
https://doi.org/10.9745/GHSP-D-20-00234

Hayes, C. N., Zhang, P., Zhang, Y., & Chayama, K. (2018).
Molecular Mechanisms of Hepatocarcinogenesis Following
Sustained Virological Response in Patients with Chronic
Hepatitis C Virus Infection. Viruses, 10(10), 531.
https://doi.org/10.3390/v10100531

Hepatitis C. (n.d.). Hepatitis C. Retrieved August 8, 2023, from
https://www.who.int/news-room/fact-sheets/detail/hepatitis-c

Interview with Wahid Doss on the 100 Million Healthy Lives
Project. (2019). Access Health International.
https://accessh.org/wp-content/uploads/2019/11/Wahid-Doss-
Final-1-1.pdf

Lingala, S., & Ghany, M. G.. (2015). Natural History of Hepatitis
C. 44(4). https://doi.org/10.1016/J.GTC.2015.07.003

Manns, M. P., McHutchison, J. G., Gordon, S. C., Rustgi, V. K.,
Shiffman, M., Reindollar, R., Goodman, Z. D., Koury, K.,
Ling, M., & Albrecht, J. K. (2001). Peginterferon alfa-2b plus
ribavirin compared with interferon alfa-2b plus ribavirin for
initial treatment of chronic hepatitis C: a randomised trial.
Lancet (London, England), 358(9286), 958–965.
https://doi.org/10.1016/s0140-6736(01)06102-5

Messina, J. P., Humphreys, I., Flaxman, A., Brown, A., Cooke, G.
S., Pybus, O. G., & Barnes, E. (2015). Global distribution and

prevalence of hepatitis C virus genotypes. Hepatology (Baltimore, Md.), 61(1), 77–87. https://doi.org/10.1002/hep.27259

Nevola, R., Rosato, V., Conturso, V., Perillo, P., Le Pera, T., Del Vecchio, F., Mastrocinque, D., Pappalardo, A., Imbriani, S., Delle Femine, A., Piacevole, A., & Claar, E. (2022). Can Telemedicine Optimize the HCV Care Cascade in People Who Use Drugs? Features of an Innovative Decentralization Model and Comparison with Other Micro-Elimination Strategies. Biology, 11(6), 805. https://doi.org/10.3390/biology11060805

Oancea, C. N., Butaru, A. E., Streba, C. T., Pirici, D., Rogoveanu, I., Diculescu, M. M., & Gheonea, D. I. (2020). Global hepatitis C elimination: history, evolution, revolutionary changes and barriers to overcome. Romanian journal of morphology and embryology = Revue roumaine de morphologie et embryologie, 61(3), 643–653. https://doi.org/10.47162/RJME.61.3.02

Offersgaard, A., Bukh, J., & Gottwein, J. M. (2023). Toward a vaccine against hepatitis C virus. Science (New York, N.Y.), 380(6640), 37–38. https://doi.org/10.1126/science.adf2226

Petruzziello, A., Marigliano, S., Loquercio, G., Cozzolino, A., & Cacciapuoti, C. (2016). Global epidemiology of hepatitis C virus infection: An up-date of the distribution and circulation of hepatitis C virus genotypes. World journal of gastroenterology, 22(34), 7824–7840. https://doi.org/10.3748/wjg.v22.i34.7824

Ren, J., Ellis, J., & Li, J. (2014). Influenza A HA's conserved epitopes and broadly neutralizing antibodies: a prediction method. Journal of bioinformatics and computational biology, 12(5), 1450023. https://doi.org/10.1142/S0219720014500231

Rongey, C. A., Kanwal, F., Hoang, T., Gifford, A. L., & Asch, S. M. (2009). Viral RNA testing in hepatitis C antibody-positive veterans. American journal of preventive medicine, 36(3), 235–238. https://doi.org/10.1016/j.amepre.2008.10.013

Sepulveda-Crespo, D., Resino, S., & Martinez, I. (2020). Hepatitis C virus vaccine design: Focus on the humoral immune response. Journal of Biomedical Science, 27(1). https://doi.org/10.1186/s12929-020-00669-4

Smith, D. B., Bukh, J., Kuiken, C., Muerhoff, A. S., Rice, C. M., Stapleton, J. T., & Simmonds, P. (2014). Expanded classification of hepatitis C virus into 7 genotypes and 67 subtypes: updated criteria and genotype assignment web resource. Hepatology (Baltimore, Md.), 59(1), 318–327. https://doi.org/10.1002/hep.26744

Tan, J. A., Joseph, T. A., & Saab, S. (2008). Treating hepatitis C in the prison population is cost-saving. Hepatology (Baltimore, Md.), 48(5), 1387–1395. https://doi.org/10.1002/hep.22509

The Global Fund. (2020, June). Global fund COVID-19 report: Deaths from HIV, TB and malaria could almost double in 12 months unless urgent action is taken. https://www.theglobalfund.org/en/news/2020/2020-06-24-global-fund-covid-19-report-deaths-from-hiv-tb-and-malaria-could-almost-double-in-12-months-unless-urgent-action-is-taken/

The Global Fund. (2023, August). Global Fund Agreements substantially reduce the price of first-line HIV treatment to below US$45 a year. https://www.theglobalfund.org/en/news/2023/2023-08-30-global-fund-agreements-substantially-reduce-price-first-line-hiv-treatment-below-usd45-a-year/

Thedja, M. D., Wibowo, D. P., El-Khobar, K. E., Ie, S. I., Turyadi, Setiawan, L., Murti, I. S., & Muljono, D. H. (2021). Improving Linkage to Care of Hepatitis C: Clinical Validation of GeneXpert® HCV Viral Load Point-of-Care Assay in Indonesia. The American journal of tropical medicine and hygiene, 105(1), 117–124. https://doi.org/10.4269/ajtmh.20-1588

Thimme R. (2021). T cell immunity to hepatitis C virus: Lessons for a prophylactic vaccine. Journal of hepatology, 74(1), 220–229. https://doi.org/10.1016/j.jhep.2020.09.022

Thompson, W. W., Symum, H., Sandul, A., DHSc, Gupta, N., Patel, P., Nelson, N., Mermin, J., & Wester, C. (2022). Vital Signs: Hepatitis C Treatment Among Insured Adults - United States, 2019-2020. MMWR. Morbidity and mortality weekly report, 71(32), 1011–1017. https://doi.org/10.15585/mmwr.mm7132e1

Thornton J. (2022). Botswana's HIV/AIDS success. Lancet (London, England), 400(10351), 480–481. https://doi.org/10.1016/S0140-6736(22)01523-9

Trepo C. (2014). A brief history of hepatitis milestones. Liver international : official journal of the International Association for the Study of the Liver, 34 Suppl 1, 29–37. https://doi.org/10.1111/liv.12409

Waked, I.. (2020). Screening and Treatment Program to Eliminate Hepatitis C in Egypt. 382(12), 1166–1174. https://doi.org/10.1056/NEJMsr1912628

Wester, C., Osinubi, A., Kaufman, H. W., Symum, H., Meyer, W. A., 3rd, Huang, X., & Thompson, W. W. (2023). Hepatitis C Virus Clearance Cascade - United States, 2013-2022. MMWR. Morbidity and mortality weekly report, 72(26), 716–720. https://doi.org/10.15585/mmwr.mm7226a3

Wijesundere, D. A., & Ramasamy, R. (2017). Analysis of Historical Trends and Recent Elimination of Malaria from Sri Lanka and Its Applicability for Malaria Control in Other Countries. Frontiers in public health, 5, 212. https://doi.org/10.3389/fpubh.2017.00212

World Bank Group. (2018). International bank for reconstruction and development project appraisal document on a proposed loan in the amount of US$530 million to the Arab Republic of Egypt for a transforming Egypt's healthcare system project. [data file] Retrieved from https://documents1.worldbank.org/curated/pt/796381530329773770/text/Egypt-PAD-06082018.txt

World Health Organization. (2017). Global hepatitis report 2017. World Health Organization.

Zein N. N. (2000). Clinical significance of hepatitis C virus genotypes. Clinical microbiology reviews, 13(2), 223–235. https://doi.org/10.1128/CMR.13.2.223

Zingaretti, C., De Francesco, R., & Abrignani, S. (2014). Why is it so difficult to develop a hepatitis C virus preventive vaccine?. Clinical microbiology and infection : the official publication of the European Society of Clinical Microbiology and Infectious Diseases, 20 Suppl 5, 103–109. https://doi.org/10.1111/1469-0691.12493